FREE BOOKS

www.*forgottenbooks*.org

You can read literally <u>thousands</u> of books for free at www.forgottenbooks.org

(please support us by visiting our web site)

Forgotten Books takes the uppermost care to preserve the entire content of the original book. However, this book has been generated from a scan of the original, and as such we cannot guarantee that it is free from errors or contains the full content of the original. But we try our best!

Truth may seem, but cannot be:
Beauty brag, but 'tis not she;
Truth and beauty buried be.

To this urn let those repair
That are either true or fair;
For these dead birds sigh a prayer.

Bacon

Harper's Library *of* Living Thought

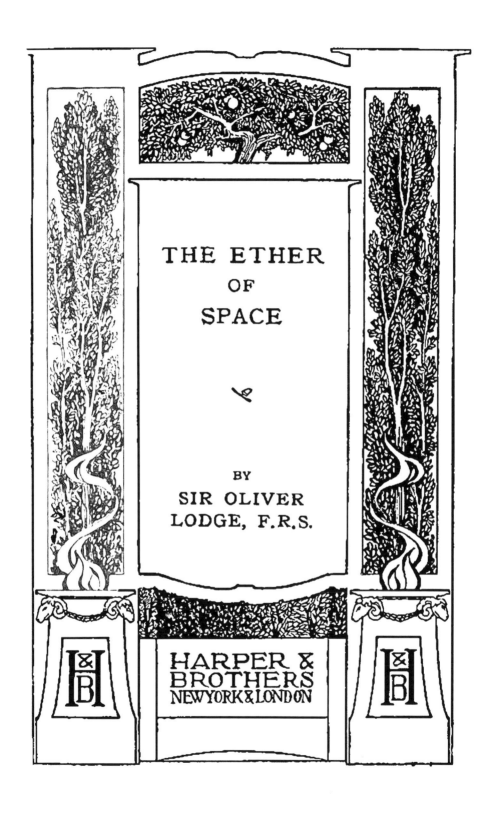

FIG. 15.—View of Ether machine complete and in action.
(See Chapter V, and Figs. 12 and 13.)

THE ETHER OF SPACE

BY

SIR OLIVER LODGE, F.R.S.

D.Sc. Lond., Hon. D.Sc. Oxon. et Vict.
LL.D. St. Andrew's, Glasgow, and Aberdeen
Vice-President of the Institution of Electrical Engineers
Rumford Medallist of the Royal Society
Ex-President of the Physical Society of London
Late Professor of Physics in the University College of Liverpool
Honorary Member of the American Philosophical Society of Philadelphia;
of the Manchester Philosophical Society; of the Batavian
Society of Rotterdam; and of the Academy of Sciences of Bologna
Principal of the University of Birmingham

ILLUSTRATED

NEW YORK AND LONDON
HARPER & BROTHERS

Copyright, 1909, by HARPER & BROTHERS.

All rights reserved.

Published May, 1909.

TO THE FOUNDERS OF

UNIVERSITY COLLEGE, LIVERPOOL,

ESPECIALLY TO THOSE BEARING THE NAMES

OF RATHBONE AND OF HOLT

THIS BOOK IS INSCRIBED

PREFACE

INVESTIGATION of the nature and properties of the Ether of Space has long been for me the most fascinating branch of Physics, and I welcome the opportunity of attempting to make generally known the conclusions to which I have so far been led on this great and perhaps inexhaustible subject.

OLIVER LODGE.

UNIVERSITY OF BIRMINGHAM,
March, 1909.

CONTENTS

CHAPTER		PAGE
	INTRODUCTION. GENERAL AND HISTORICAL	xv
I.	THE LUMINIFEROUS ETHER AND THE MODERN THEORY OF LIGHT	1
II.	THE INTERSTELLAR ETHER AS A CONNECTING MEDIUM	13
III.	INFLUENCE OF MOTION ON VARIOUS PHENOMENA	30
IV.	EXPERIMENTS ON THE ETHER	46
V.	SPECIAL EXPERIMENT ON ETHERIAL VISCOSITY	70
VI.	ETHERIAL DENSITY	88
VII.	FURTHER EXPLANATIONS CONCERNING THE DENSITY AND ENERGY OF THE ETHER	95
VIII.	ETHER AND MATTER	107
IX.	STRENGTH OF THE ETHER	124
X.	GENERAL THEORY OF ABERRATION . .	136
APPENDIX 1.	ON GRAVITY AND ETHERIAL TENSION.	153
APPENDIX 2.	CALCULATIONS IN CONNECTION WITH ETHER DENSITY.	156
APPENDIX 3.	FRESNEL'S LAW A SPECIAL CASE OF A UNIVERSAL POTENTIAL FUNCTION	163

LIST OF ILLUSTRATIONS

Illustrations of Aberration

FIG.		PAGE
1.	Cannon shots	36
2.	Boats or Waves	37
3.	Lighthouse beams	38
4.	Ray through a moving stratum	41
5.	Wave-fronts in moving medium	43
6.	Normal reflection in moving medium	44

Experiments on Ether drift

7.	Interference Kaleidoscope	53
8.	Hoek's experiment	56
9.	Experiment of Mascart and Jamin	57
10.	Diagram of Michelson's experiment	64

Illustrations of Ether Machine (Lodge)

11.	Diagram of course of light	72
12.	General view of whirling part of Ether Machine	76
13.	General view of optical frame	79
14.	Drawing of optical details	*Facing p.* 80
15.	View of Ether Machine in action	*Frontispiece*
16.	Appearance of interference bands and micrometer wires	80
17.	Iron mass for magnetisation	84
18.	Appearance of bands	83
19.	Arrangement for electrification	85

INTRODUCTION

"ETHER or Æther ($\alpha i\theta \acute{\eta}\rho$ probably from $\alpha i\theta\omega$ I burn,) a material substance of a more subtle kind than visible bodies, supposed to exist in those parts of space which are apparently empty."

So begins the article, "Ether," written for the ninth edition of the *Encyclopædia Britannica*, by James Clerk-Maxwell.

The derivation of the word seems to indicate some connection in men's minds with the idea of Fire: the other three "elements," Earth, Water, Air, representing the solid, liquid, and gaseous conditions of ordinary matter respectively. The name Æther suggests a far more subtle or penetrating and ultra-material kind of substance.

Newton employs the term for the medium which fills space—not only space which appears to be empty, but space also which appears to be full; for the luminiferous ether must undoubtedly

INTRODUCTION

penetrate between the atoms—must exist in the pores so to speak—of every transparent substance, else light could not travel through it. The following is an extract from Newton's surmises concerning this medium:—

"Qu. 18. If in two large tall cylindrical Vessels of Glass inverted, two little Thermometers be suspended so as not to touch the Vessels, and the Air be drawn out of one of these Vessels, and these Vessels thus prepared be carried out of a cold place into a warm one; the Thermometer *in vacuo* will grow warm as much and almost as soon as the Thermometer which is not *in vacuo*. And when the Vessels are carried back into the cold place, the Thermometer *in vacuo* will grow cold almost as soon as the other Thermometer. Is not the Heat of the warm Room conveyed through the Vacuum by the Vibrations of a much subtiler Medium than Air, which after the Air was drawn out remained in the Vacuum? And is not this Medium the same with that Medium by which Light is [transmitted], and by whose

INTRODUCTION

Vibrations Light communicates Heat to Bodies? ... And do not the Vibrations of this Medium in hot Bodies contribute to the intenseness and duration of their Heat? And do not hot Bodies communicate their Heat to contiguous cold ones by the Vibrations of this Medium propagated from them into the cold ones? And is not this Medium exceedingly more rare and subtile than the Air, and exceedingly more elastick and active? And doth it not readily pervade all bodies? And is it not (by its elastick force) expanded through all the Heavens?"

"Qu. 22. May not Planets and Comets, and all gross Bodies, perform their motions more freely, and with less resistance in this Æthereal Medium than in any Fluid, which fills all Space adequately without leaving any Pores, and by consequence is much denser than Quick-silver and Gold? And may not its resistance be so small, as to be inconsiderable? For instance; if this *Æther* (for so I will call it) should be supposed 700000 times more elastick than our Air, and above 700000 times more rare; its

INTRODUCTION

resistance would be above 600,000,000 times less than that of Water. And so small a resistance would scarce make any sensible alteration in the Motions of the Planets in ten thousand Years."

That the ether, if there be such a thing in space, can pass readily into or through matter is often held proven by tilting a mercury barometer; when the mercury rises to fill the transparent vacuum. Everything points to its universal permeance, if it exist at all.

But these, after all, are antique thoughts. Electric and Magnetic information has led us beyond them into a region of greater certainty and knowledge; so that now I am able to advocate a view of the Ether which makes it not only uniformly present and all-pervading, but also massive and substantial beyond conception. It is turning out to be by far the most substantial thing—perhaps the only substantial thing—in the material universe. Compared to ether the densest matter, such as lead or gold, is a filmy gossamer structure; like a comet's tail or a milky way, or like a salt in very dilute solution.

INTRODUCTION

To lead up to and justify the idea of the reality and substantiality, and vast though as yet largely unrecognized importance, of the Ether of Space, the following chapters have been written. Some of them represent the expanded notes of lectures which have been given in various places —chiefly the Royal Institution; while the first chapter represents a lecture before the Ashmolean Society of the University of Oxford in June, 1889. One chapter (*viz.*, Chap. II) has already been printed as part of an appendix to the third edition of *Modern Views of Electricity*, as well as in the *Fortnightly* and *North American Reviews;* but no other chapters have yet been published, though parts appear in more elaborate form in Proceedings or Transactions of learned societies.

The problem of the constitution of the Ether, and of the way in which portions of it are modified to form the atoms or other constituent units of ordinary matter, has not yet been solved. Much work has been done in this direction by various mathematicians, but much more remains to be done. And until it is done, some scepticism is reasonable—perhaps laudable. Meanwhile there are few physicists who will

INTRODUCTION

dissent from Clerk-Maxwell's penultimate sentence in the article "Ether," of which the beginning has already been quoted:—

"Whatever difficulties we may have in forming a consistent idea of the constitution of the æther, there can be no doubt that the interplanetary and interstellar spaces are not empty, but are occupied by a material substance or body, which is certainly the largest, and probably the most uniform body of which we have any knowledge."

THE ETHER OF SPACE

THE ETHER OF SPACE

I

THE LUMINIFEROUS ETHER AND THE MODERN THEORY OF LIGHT

THE oldest and best known function for an ether is the conveyance of light, and hence the name "luminiferous" was applied to it; though at the present day many more functions are known, and more will almost certainly be discovered.

To begin with, it is best to learn what we can concerning the properties of the Interstellar Ether from the phenomena of Light.

For now well-nigh a century we have had a wave theory of light; and a wave theory of light is quite certainly true. It is directly demonstrable that light consists of waves of some kind or other, and that these waves travel at a certain well-known velocity, achieving a distance equal to seven times the circumference of the earth every second; from New York to London

THE ETHER OF SPACE

and back in the thirtieth part of a second; and taking only eight minutes on the journey from the sun to the earth. This propagation in time of an undulatory disturbance necessarily involves a medium. If waves setting out from the sun exist in space eight minutes before striking our eyes, there must necessarily be in space some medium in which they exist and which conveys them. Waves we cannot have, unless they be waves in something.

No ordinary matter is competent to transmit waves at anything like the speed of light: the rate at which *matter* conveys waves is the velocity of sound—a speed comparable to one-millionth of the speed of light. Hence the luminiferous medium must be a special kind of substance; and it is called the ether. The *luminiferous* ether it used to be called, because the conveyance of light was all it was then known to be capable of; but now that it is known to do a variety of other things also, the qualifying adjective may be dropped. But, inasmuch as the term "ether" is also applied to a familiar organic compound, we may distinguish the ultramaterial luminiferous medium by calling it the Ether of Space.

Wave motion in ether, light certainly is; but what does one mean by the term wave? The popular notion is, I suppose, of something heaving up and down, or perhaps of something break-

THEORY OF LIGHT

ing on a shore. But if you ask a mathematician what he means by a wave, he will probably reply that the most general wave is such a function of x and y and t as to satisfy the differential equation

$$\frac{d^2y}{dt^2} = v^2 \frac{d^2y}{dx^2};$$

while the simplest wave is

$$y = a \sin (x - vt).$$

And he might possibly refuse to give any other answer.

And in refusing to give any other answer than this, or its equivalent in ordinary words, he is entirely justified; that *is* what is meant by the term wave, and nothing less general would be all-inclusive.

Translated into ordinary English, the phrase signifies, with accuracy and comprehensive completeness, the full details of "a disturbance periodic both in space and time." Anything thus doubly periodic is a wave; and all waves—whether in air as sound waves, or in ether as light waves, or on the surface of water as ocean waves—can be comprehended in the definition.

What properties are essential to a medium capable of transmitting wave motion? Roughly, we may say two: *elasticity* and *inertia*. Elasticity in some form, or some equivalent of it, in order

THE ETHER OF SPACE

to be able to store up energy and effect recoil; inertia, in order to enable the disturbed substance to overshoot the mark and oscillate beyond its place of equilibrium to and fro. Any medium possessing these two properties can transmit waves, and unless a medium possesses these properties in some form or other, or some equivalent for them, it may be said with moderate security to be incompetent to transmit waves. But if we make this latter statement, one must be prepared to extend to the terms elasticity and inertia their very largest and broadest signification, so as to include any possible kind of restoring force, and any possible kind of persistence of motion, respectively.

These matters may be illustrated in many ways, but perhaps a simple loaded lath, or spring, in a vise will serve well enough. Pull it to one side, and its elasticity tends to make it recoil; let it go, and its inertia causes it to overshoot its normal position. That is what inertia is: power of overshooting a mark, or, more accurately, power of moving for a time even against driving force—power to rush up hill. Both causes together make it swing to and fro till its energy is exhausted. This is a disturbance simply periodic in time. A regular series of such springs, set at equal intervals and started vibrating at regular intervals of time one after the other, would be periodic in space too; and

THEORY OF LIGHT

so they would, in disconnected fashion, typify a wave. A series of pendulums will do just as well, and if set swinging in orderly fashion will furnish at once an example and an appearance of wave motion which the most casual observer must recognise as such. The row of springs obviously possesses elasticity and inertia; and any wave-transmitting medium must similarly possess some form of elasticity and some form of inertia.

But now proceed to ask what is this Ether which in the case of light is thus vibrating? What corresponds to the elastic displacement and recoil of the spring or pendulum? What corresponds to the inertia whereby it overshoots its mark? Do we know these properties in the ether in any other way?

The answer, given first by Clerk-Maxwell, and now reiterated and insisted on by experiments performed in every important laboratory in the world, is:—

> The elastic displacement corresponds to electrostatic charge—roughly speaking, to electricity.
> The inertia corresponds to magnetism.

This is the basis of the modern electromagnetic theory of light.

Let me attempt to illustrate the meaning of this statement, by reviewing some fundamental electrical facts in the light of these analogies:—

THE ETHER OF SPACE

The old and familiar operation of charging a Leyden jar — the storing up of energy in a strained dielectric—any electrostatic charging whatever—is quite analogous to the drawing aside of our flexible spring. It is making use of the elasticity of the ether to produce a tendency to recoil. Letting go the spring is analogous to permitting a discharge of the jar—permitting the strained dielectric to recover itself—the electrostatic disturbance to subside.

In nearly all the experiments of electrostatics etherial elasticity is manifest.

Next consider inertia. How would one illustrate the fact that water, for instance, possesses inertia—the power of persisting in motion against obstacles—the power of possessing kinetic energy? The most direct way would be to take a stream of water and try suddenly to stop it. Open a water-tap freely and then suddenly shut it. The impetus or momentum of the stopped water makes itself manifest by a violent shock to the pipe, with which everybody must be familiar. This momentum of water is utilised by engineers in the "water-ram."

A precisely analogous experiment in Electricity is what Faraday ca led "the extra current." Send a current through a coil of wire round a piece of iron, or take any other arrangement for developing powerful magnetism, and then suddenly stop the current by breaking the circuit.

THEORY OF LIGHT

A violent flash occurs if the stoppage is sudden enough—a flash which means the bursting of the insulating air partition by the accumulated electromagnetic momentum. The scientific name for this electrical inertia is "self-induction."

Briefly we may say that nearly all electromagnetic experiments illustrate the fact of etherial inertia.

Now return to consider what happens when a charged conductor (say a Leyden jar) is discharged. The recoil of the strained dielectric causes a current, the inertia of this current causes it to overshoot the mark, and for an instant the charge of the jar is reversed; the current now flows backward and charges the jar up as at first; back again flows the current; and so on, charging and reversing the charge, with rapid oscillations, until the energy is all dissipated into heat. The operation is precisely analogous to the release of a strained spring, or to the plucking of a stretched string.

But the discharging body, thus thrown into strong electrical vibration, is imbedded in the all-pervading ether; and we have just seen that the ether possesses the two properties requisite for the generation and transmission of waves—*viz.*, elasticity, and inertia or density; hence, just as a tuning-fork vibrating in air excites aerial waves, or sound, so a discharging Leyden jar in ether excites etherial waves, or light.

THE ETHER OF SPACE

Etherial waves can, therefore, be actually produced by direct electrical means. I discharge here a jar, and the room is for an instant filled with light. With light, I say, though you can see nothing. You can see and hear the spark, indeed; but that is a mere secondary disturbance we can for the present ignore—I do not mean any secondary disturbance. I mean the true etherial waves emitted by the electric oscillation going on in the neighbourhood of the recoiling dielectric. You pull aside the prong of a tuning-fork and let it go: vibration follows and sound is produced. You charge a Leyden jar and let it discharge: vibration follows and light is excited.

It is light, just as good as any other light. It travels at the same pace, it is reflected and refracted according to the same laws; every experiment known to optics can be performed with this etherial radiation electrically produced—and yet you cannot see it. Why not? For no fault of the light; the fault (if there be a fault) is in the eye. The retina is incompetent to respond to these vibrations—they are too slow. The vibrations set up when this large jar is discharged are from a hundred thousand to a million per second, but that is too slow for the retina. It responds only to vibrations between 400 billion and 700 billion per second. The vibrations are too quick for the ear, which re-

8

THEORY OF LIGHT

sponds only to vibrations between 40 and 40,000 per second. Between the highest audible and the lowest visible vibrations there has been hitherto a great gap, which these electric oscillations go far to fill up. There has been a great gap simply because we have no intermediate sense organ to detect rates of vibration between 40,000 and 400,000,000,000,000 per second. It was therefore an unexplored territory. Waves have been there all the time in any quantity, but we have not thought about them nor attended to them.

It happens that I have myself succeeded in getting electric oscillations so slow as to be audible—the lowest I had got in 1889 were 125 per second, and for some way above this the sparks emit a musical note; but no one has yet succeeded in directly making electric oscillations which are visible—though indirectly everyone does it when they light a candle.

It is easy, however, to have an electric oscillator which vibrates 300 million times a second, and emits etherial waves a yard long. The whole range of vibrations between musical tones and some thousand million per second is now filled up.

With the large condensers and self-inductances employed in modern cable telegraphy, it is easy to get a series of beautifully regular and gradually damped electric oscillations, with a period of

THE ETHER OF SPACE

two or three seconds, recorded by an ordinary signalling instrument or siphon recorder.

These electromagnetic waves in space have been known on the side of theory ever since 1865, but interest in them was immensely quickened by the discovery of a receiver or detector for them. The great though simple discovery by Hertz, in 1888, of an "electric eye," as Lord Kelvin called it, made experiments on these waves for the first time easy or even possible. From that time onward we possessed a sort of artificial sense organ for their appreciation — an electric arrangement which can virtually "see" these intermediate rates of vibration.

Since then Branly discovered that metallic powder could be used as an extraordinarily sensitive detector; and on the basis of this discovery, the "coherer" was employed by me for distant signalling by means of electric or etheric waves, until now when many other detectors are available in the various systems of wireless telegraphy.

With these Hertzian waves all manner of optical experiments can be performed. They can be reflected by plain sheets of metal, concentrated by parabolic reflectors, refracted by prisms, and concentrated by lenses. I have made, for instance, a large lens of pitch, weighing over three hundredweight, for concentrating them to a focus.[1] They can be made to show the phenomenon of interference, and thus have their wave-length accurately measured. They are stopped by all conductors, and transmitted by all insulators. Metals are opaque; but even imperfect insulators, such as wood or stone, are strikingly transparent; and waves may be received in one room from a source in another, the door between the two being shut.

The real nature of metallic opacity and of transparency has long been clear in Maxwell's theory of light, and these electrically produced waves only illustrate and bring home the well-known facts. The experiments of Hertz are, in fact, the apotheosis of Maxwell's theory.

Thus, then, in every way, Clerk-Maxwell's brilliant perception or mathematical deduction, in 1865, of the real nature of light is abundantly justified; and for the first time we have a true theory of light—no longer based upon analogy with sound, nor upon the supposed properties of some hypothetical jelly or elastic solid, but capable of being treated upon a substantial basis of its own, in alliance with the sciences of Electricity and of Magnetism.

Light is an electromagnetic disturbance of the ether. Optics is a branch of electricity. Out-

[1] See Lodge and Howard, *Philosophical Magazine* for July, 1889. See also *Phil. Mag.*, August, 1888, page 229.

THE ETHER OF SPACE

standing problems in optics are being rapidly solved, now that we have the means of definitely exciting light with a full perception of what we are doing, and of the precise mode of its vibration.

It remains to find out how to shorten down the waves—to hurry up the vibration until the light becomes visible. Nothing is wanted but quicker modes of vibration. Smaller oscillators must be used—very much smaller—oscillators not much bigger than molecules. In all probability—one may almost say certainly—ordinary light is the result of electric oscillation in the molecules or atoms of hot bodies, or sometimes of bodies not hot—as in the phenomenon of phosphorescence.

The direct generation of *visible* light by electric means, so soon as we have learnt how to attain the necessary frequency of vibration, will have most important practical consequences; and that matter is initially dealt with in a section on the Manufacture of Light, § 149, in Chapter XIV of *Modern Views of Electricity*. But here we abandon further consideration of this aspect of our great subject.

II

THE INTERSTELLAR ETHER AS A CONNECTING MEDIUM

SO far I have given a general idea of the present condition of the wave theory of light, both from its theoretical and from its experimental sides. The waves of light are not anything mechanical or material, but are something electrical and magnetic—they are, in fact, electrical disturbances periodic in space and time, and travelling with a known and tremendous speed through the ether of space. Their very existence depends upon the ether, and their speed of propagation is its best known and most certain quantitative property.

A statement of this kind does not even initially express a tithe of our knowledge on the subject; nor does our knowledge exhaust any large part of the region of discoverable fact; but the statement above made may be regarded as certain, although the absence of mechanics or ordinary dynamics about it removes it, or seems to remove it, from the category of the historically soundest

THE ETHER OF SPACE

and best worked department of Physical Science —*viz.*, that explored by the Newtonian method. Though in truth there is every reason to suppose that we should have had Newton with us in these modern developments.

There is, I believe, a general tendency to underrate the certainty of some of the convictions to which natural philosophers have gradually, in the course of their study of nature, been impelled; more especially when those convictions have reference to something intangible and occult. The existence of a continuous space-filling medium, for instance, is probably regarded by most educated people as a more or less fanciful hypothesis, a figment of the scientific imagination—a mode of collating and welding together a certain number of observed facts, but not in any physical sense a reality, as water and air are realities.

I am speaking purely physically. There may be another point of view from which all material reality can be denied, but with those questions physics proper has nothing to do; it accepts the evidence of the senses, regarding them as the tools or instruments wherewith man may hope to understand one definite aspect of the universe; and it leaves to philosophers, equipped from a different armory, the other aspects which the material universe may—nay, must—possess.

By a physical "explanation" is meant a clear

A CONNECTING MEDIUM

statement of a fact or law in terms of something with which daily life has made us familiar. We are all chiefly familiar, from our youth up, with two apparently simple things, *motion* and *force*. We have a direct sense for both these things. We do not understand them in any deep way, probably we do not understand them at all, but we are accustomed to them. Motion and force are our primary objects of experience and consciousness; and in terms of them all other less familiar occurrences may conceivably be stated and grasped. Whenever a thing can be so clearly and definitely stated, it is said to be explained, or understood; we are said to have "a dynamical theory" of it. Anything short of this may be a provisional or partial theory, an explanation of the less known in terms of the more known, but Motion and Force are postulated in physics as the completely known: and no attempt is made to press the terms of an explanation further than that. A dynamical theory is recognized as being at once necessary and sufficient.

Now, it must be admitted at once that of very few things have we at present such a dynamical explanation. We have no such explanation of matter, for instance, or of gravitation, or of electricity, or ether, or light. It is always conceivable that of some such things no purely dynamical explanation will ever be forthcoming,

THE ETHER OF SPACE

because something more than motion and force may perhaps be essentially involved. Still, physics is bound to push the search for an explanation to its furthest limits; and so long as it does not hoodwink itself by vagueness and mere phrases—a feebleness against which its leaders are mightily and sometimes cruelly on their guard, preferring to risk the rejection of worthy ideas rather than permit a semi-acceptance of anything fanciful and obscure—so long as it vigorously probes all phenomena within its reach, seeking to reduce the physical aspect of them to terms of motion and force—so long it must be upon a safe track. And, by its failure to deal with certain phenomena, it will learn—it already begins to suspect, its leaders must long have surmised—the existence of some third, as yet unknown, category, by incorporating which the physics of the future may rise to higher flights and an enlarged scope.

I have said that the things of which we are permanently conscious are motion and force, but there is a third thing which we have likewise been all our lives in contact with, and which we know even more primarily, though perhaps we are so immersed in it that our knowledge realises itself later—*viz.*, life and mind. I do not now pretend to define these terms, or to speculate as to whether the things they denote are essentially one and not two. They *exist*, in the sense in

A CONNECTING MEDIUM

which we permit ourselves to use that word, and they are not yet incorporated into physics. Till they are, they may remain more or less vague; but how or when they can be incorporated, is not for me even to conjecture.

Still, it is open to a physicist to state how the universe appears to him, in its broad character and physical aspect. If I were to make the attempt, I should find it necessary, for the sake of clearness, to begin with the simplest and most fundamental ideas; in order to illustrate, by facts and notions in universal knowledge, the kind of process which essentially occurs in connection with the formation of higher and less familiar conceptions—in regions where the common information of the race is so slight as to be useless.

Primary Acquaintance with the External World.

Beginning with our most fundamental sense, I should sketch the matter thus:—

We have muscles and can move. I cannot analyze motion—I doubt if the attempt is wise—it is a simple immediate act of perception, a direct sense of free unresisted muscular action. We may indeed move without feeling it, and that teaches us nothing; but we may move so as to feel it, and this teaches us much, and leads to our first scientific inference—*viz., space;* that is, simply, room to move about. We might

THE ETHER OF SPACE

have had a sense of being jammed into a full or tight-packed universe; but we have not: we feel it to be a spacious one.

Of course we do not stop at this baldness of inference: our educated faculty leads us to realise the existence of space far beyond the possibility of direct sensation; and, further, by means of the direct appreciation of *speed* in connection with motion—of uniform and variable speed—we become able to formulate the idea of "time," or uniformity of sequence; and we attain other more complex notions—acceleration, and the like—upon a consideration of which we need not now enter.

But our muscular sense is not limited to the perception of free motion: we constantly find it restricted or forcibly resisted. This "muscular action impeded" is another direct sense, that of *"force"*; and attempts to analyze it into anything simpler than itself have hitherto resulted only in confusion. By "force" is meant primarily muscular action not accompanied by motion. Our sense of this teaches us that space, though roomy, is not empty: it gives us our second scientific inference—what we call "matter."

Again we do not stop at this bare inference. By another sense, that of pain, or mere sensation, we discriminate between masses of matter in apparently intimate relation with ourselves,

A CONNECTING MEDIUM

and other or foreign lumps of matter; and we use the first portion as a measure of the extent of the second. The human body is our standard of size. We proceed also to subdivide our idea of matter—according to the varieties of resistance with which it appeals to our muscular sense—into four different states, or "elements," as the ancients called them—*viz.*, the solid, the liquid, the gaseous, and the ethereal. The resistance experienced when we encounter one or other of these forms of material existence varies from something very impressive—the solid ; through something nearly impalpable—the gaseous ; up to something entirely imaginative, fanciful, or inferential—*viz.*, the ether.

The ether does not in any way affect our sense of touch (*i.e.*, of force); it does not resist motion in the slightest degree. Not only can our bodies move through it, but much larger bodies, planets and comets, can rush through it at what we are pleased to call a prodigious speed (being far greater than that of an athlete) without showing the least sign of friction. I myself, indeed, have designed and carried out a series of delicate experiments to see whether a whirling mass of iron could to the smallest extent grip the ether and carry it round, with so much as a thousandth part of its own velocity. These shall be described further on, but meanwhile the result arrived at is distinct. The answer is, no; I

THE ETHER OF SPACE

cannot find a trace of mechanical connection between matter and ether, of the kind known as viscosity or friction.

Why, then, if it is so impalpable, should we assert its existence? May it not be a mere fanciful speculation, to be extruded from physics as soon as possible? If we were limited for our knowledge of matter to our sense of touch, the question would never even have presented itself; we should have been simply ignorant of the ether, as ignorant as we are of any life or mind in the universe not associated with some kind of material body. But our senses have attained a higher stage of development than that. We are conscious of matter by means other than its resisting force. Matter acts on one small portion of our body in a totally different way, and we are said to *taste* it. Even from a distance it is able to fling off small particles of itself sufficient to affect another delicate sense. Or again, if it is vibrating with an appropriate frequency, another part of our body responds; and the universe is discovered to be not silent but eloquent to those who have ears to hear. Are there any more discoveries to be made? Yes; and already some have been made. All the senses hitherto mentioned speak to us of the presence of ordinary matter—gross matter, as it is sometimes called—though when appealing to our sense of smell, and more especially to a dog's sense of

A CONNECTING MEDIUM

smell, it is not very gross; still, with the senses hitherto enumerated we should never have become aware of the ether. A stroke of lightning might have smitten our bodies back into their inorganic constituents, or a torpedo-fish might have inflicted on us a strange kind of torment; but from these violent tutors we should have learnt little more than a school-boy learns from the once ever-ready cane.

But it so happens that the whole surface of our skin is sensitive in yet another way, and a small portion of it is astoundingly and beautifully sensitive, to an impression of an altogether different character—one not necessarily associated with any form of ordinary matter—one that will occur equally well through space from which all solid, liquid, or gaseous matter has been removed. Hold your hand near a fire, put your face in the sunshine, and what is it you feel? You are now conscious of something not arriving by ordinary matter at all. You are now as directly conscious as you can be of the ethereal medium. True the process is not very direct. You cannot apprehend the ether as you can matter, by touching or tasting or even smelling it; but the process is analogous to the kind of perception we might get of ordinary matter if we had the sense of hearing alone. It is something akin to *vibrations in the ether* that our skin and our eyes feel.

It may be rightly asserted that it is not the

THE ETHER OF SPACE

ethereal disturbances themselves, but other disturbances excited by them in our tissues, that our heat nerves feel; and the same assertion can be made for our more highly developed and specialised sight nerves. All nerves must feel what is occurring next door to them, and can directly feel nothing else; but the "radiation," the cause which excited these disturbances, travelled through the ether—not through any otherwise known material substance.

It should be a commonplace to rehearse how we know this. Briefly, thus: Radiation conspicuously comes to us from the sun. If any free or ordinary matter exists in the intervening space, it must be an exceedingly rare gas. In other words, it must consist of scattered particles of matter, some big enough to be called lumps, some so small as to be merely atoms, but each with a considerable gap between it and its neighbor. Such isolated particles are absolutely incompetent to transmit light. And, parenthetically, I may say that no form of ordinary matter, solid, liquid, or gaseous, is competent to transmit a thing travelling with the speed and subject to the known laws of light. For the conveyance of radiation or light all ordinary matter is not only incompetent, but hopelessly and absurdly incompetent. If this radiation is a thing transmitted by anything at all, it must be by something *sui generis*.

A CONNECTING MEDIUM

But it is transmitted; for it takes time on the journey, travelling at a well-known and definite speed; and it is a quivering or periodic disturbance, falling under the general category of wave-motion. Nothing is more certain than that. No physicist disputes it. Newton himself, who is commonly and truly asserted to have promulgated a rival theory, felt the necessity of an ethereal medium, and knew that light consisted essentially of waves.

Sight.

A small digression here, to avoid any possible confusion due to the fact that I have purposely associated together temperature nerves and sight nerves. They are admittedly not the same, but they are alike in this, that they both afford evidence of radiation; and, were we blind, we might still know a good deal about the sun, and if our temperature nerves were immensely increased in delicacy (not all over, for that would be merely painful, but in some protected region), we might even learn about the moon, planets, and stars. In fact, an eye, consisting of a pupil (preferably a lens) and a sunken cavity lined with a surface sensitive to heat, could readily be imagined, and might be somewhat singularly effective. It would be more than a light recorder; it could detect all the ethereal quiverings caused

THE ETHER OF SPACE

by surrounding objects, and hence would see perfectly well in what we call "the dark." But it would probably see far too much for convenience, since it would necessarily be affected by every kind of radiation in simple proportion to its energy; unless, indeed, it were provided with a supply of screens with suitably selected absorbing powers. But whatever might be the advantage or disadvantage of such a sense-organ, we as yet do not possess one. Our eye does not act by detecting heat; in other words, it is not affected by the whole range of ethereal quiverings, but only by a very minute and apparently insignificant portion. It wholly ignores the ether waves whose frequency is comparable with that of sound; and, for thirty or forty octaves above this, nothing about us responds; but high up, in a range of vibration of the inconceivably high pitch of four to seven hundred million million per second—a range which extremely few accessible bodies are able to emit, and which it requires some knowledge and skill artificially to produce—to those waves the eye is acutely, surpassingly, and most intelligently sensitive.

This little fragment of total radiation is in itself trivial and negligible. Were it not for men, and glow-worms, and a few other forms of life, hardly any of it would ever occur, on such a moderate-sized lump of matter as the earth.

A CONNECTING MEDIUM

Except for an occasional volcano, or a flash of lightning, only gigantic bodies like the sun and stars have energy enough to produce these higher flute-like notes; and they do it by sheer main force and violence—the violence of their gravitative energy—producing not only these, but every other kind of radiation also. Glow-worms, so far as I know, alone have learned the secret of emitting the physiologically useful waves, and none others.

Why these waves are physiologically useful—why they are what is called "light," while other kinds of radiation are "dark," is a question to be asked, but, at present, only tentatively answered. The answer must ultimately be given by the Physiologist; for the distinction between light and non-light can only be stated in terms of the eye, and its peculiar specialised sensitiveness; but a hint may be given him by the Physicist. The ethereal waves which affect the eye and the photographic plate are of a size not wholly incomparable with that of the atoms of matter. When a physical phenomenon is concerned with the ultimate atoms of matter, it is often relegated at present to the field of knowledge summarized under the head of Chemistry. Sight is probably a chemical sense. The retina may contain complex aggregations of atoms, shaken asunder by the incident light vibrations, and rapidly built up again by the living tissues

THE ETHER OF SPACE

in which they live; the nerve endings meanwhile appreciating them in their temporarily dissociated condition. A vague speculation! Not to be further countenanced except as a working hypothesis leading to examination of fact; but, nevertheless, the direction in which the thoughts of some physicists are tending—a direction toward which many recently discovered experimental facts point.[1]

Gravitation and Cohesion.

It would take too long to do more than suggest some other functions for which a continuous medium of communication is necessary. We shall argue in Chapter VIII that technical action at a distance is impossible. A body can only act immediately on what it is in contact with; it must be by the action of contiguous particles —that is, practically, through a continuous medium, that force can be transmitted across space. Radiation is not the only thing the earth feels from the sun; there is in addition its gigantic gravitative pull, *a force or tension more than what a million million steel rods, each seventeen feet in diameter, could stand* (see Chap. IX). What mechanism transmits this gigantic force? Again, take a steel bar itself: when violently

[1] Cf. sections 157A, 143, 187, and chap. xvi., of my *Modern Views of Electricity.*

A CONNECTING MEDIUM

stretched, with how great tenacity its parts cling together! Yet its particles are not in absolute contact, they are only virtually attached to each other by means of the universal connecting medium—the ether—a medium that must be competent to transmit the greatest stresses which our knowledge of gravitation and of cohesion shows us to exist.

Electricity and Magnetism.

Hitherto I have mainly confined myself to the perception of the ether by our ancient sense of radiation, whereby we detect its subtle and delicate quiverings. But we are growing a new sense; not perhaps an actual sense-organ, though not so very unlike a new sense-organ, though the portions of matter which go to make the organ are not associated with our bodies by the usual links of pain and disease; they are more analogous to artificial teeth or mechanical limbs, and can be bought at an instrument-maker's.

Electroscopes, galvanometers, telephones— delicate instruments these; not yet eclipsing our sense-organs of flesh, but in a few cases coming within measurable distance of their surprising sensitiveness. And with these what do we do? Can we smell the ether, or touch it, or what is the closest analogy? Perhaps there is no useful analogy; but nevertheless we deal with it, and

THE ETHER OF SPACE

that closely. Not yet do we fully realise what we are doing. Not yet have we any dynamical theory of electric currents, of static charges, and of magnetism. Not yet, indeed, have we any dynamical theory of light. In fact, the ether has not yet been brought under the domain of simple mechanics—it has not yet been reduced to motion and force: and that probably because the *force* aspect of it has been so singularly elusive that it is a question whether we ought to think of it as material at all. No, it is apart from mechanics at present. Conceivably it may remain apart; and our first additional category, wherewith the foundations of physics must some day be enlarged, may turn out to be an ethereal one. And some such inclusion may have to be made before we can attempt to annex vital or mental processes. Perhaps they will all come in together.

Howsoever these things be, this is the kind of meaning lurking in the phrase that we do not yet know what electricity or what the ether is. We have as yet no dynamical explanation of either of them; but the past century has taught us what seems to their student an overwhelming quantity of facts about them. And when the present century, or the century after, lets us deeper into their secrets, and into the secrets of some other phenomena now in course of being rationally investigated, I feel as if it would be

A CONNECTING MEDIUM

no merely material prospect that will be opening on our view, but some glimpse into a region of the universe which Science has never entered yet, but which has been sought from far, and perhaps blindly apprehended, by painter and poet, by philosopher and saint.

Note on the Spelling of Ethereal.

The usual word "ethereal" suggests something unsubstantial, and is so used in poetry; but for the prosaic treatment of Physics it is unsuitable, and etheric has occasionally been used instead. No just derivation can be given for such an adjective, however; and I have been accustomed simply to spell etherial with an *i* when no poetic meaning was intended. This alternative spelling is not incorrect; but Milton uses the variant "ethereous," in a sense suggestive of something strong and substantial (*Par. Lost*, vi, 473). Either word, therefore, can be employed to replace "ethereal" in physics: and in succeeding chapters one or other of these is for the most part employed.

III

INFLUENCE OF MOTION ON VARIOUS PHENOMENA

NOTWITHSTANDING its genuine physical nature and properties, the ether is singularly intangible and inaccessible to our senses, and accordingly is a subject on which it is extremely difficult to try experiments. Many have been the attempts to detect some phenomena depending on its motion relative to the earth. The earth is travelling round the sun at the rate of 19 miles a second, and although this is slow compared with light—being, in fact, just about $\frac{1}{10,000}$th of the speed of light—yet it would seem feasible to observe some modification of optical phenomena due to this motion through the ether.

And one such phenomenon is indeed known—namely, the stellar aberration discovered by Bradley in 1729. The position of objects not on the earth, and not connected with the solar system, is apparently altered by an amount comparable to one part in ten thousand, by the earth's motion; that is to say, the apparent place

INFLUENCE OF MOTION

of a star is shifted from its true place by an angle $\frac{1}{10,000}$th of a "radian,"[1] or about 20 seconds of arc.

This is called Astronomical Aberration, and is extremely well known. But a number of other problems open out in connection with it, and on these it is desirable to enter into detail. For if the ether is stationary while the earth is flying through it—at a speed vastly faster than any cannon-ball, as much faster than a cannon-ball as an express train is faster than a saunter on foot—it is for all practical purposes the same as if the earth were stationary and the ether streaming past it with this immense velocity in the opposite direction. And some consequence of such a drift might at first sight certainly be expected. It might, for instance, seem doubtful whether terrestrial surveying operations can be conducted, with the extreme accuracy expected of them, without some allowance for the violent rush of the light-conveying medium past and through the theodolite of the observer.

Let us therefore consider the whole subject further.

ABERRATION.

Everybody knows that to shoot a bird on the wing you must aim in front of it. Every one will

[1] *Radian* is the name given by Prof. James Thomson to a unit angle of circular measure, an angle whose arc equals its radius, or about 57

THE ETHER OF SPACE

readily admit that to hit a squatting rabbit from a moving train you must aim behind it.

These are examples of what may be called "aberration" from the sender's point of view, from the point of view of the source. And the aberration, or needful divergence, between the point aimed at and the thing hit has opposite sign in the two cases—the case when receiver is moving, and the case when source is moving. Hence, if both be moving, it is possible for the two aberrations to neutralize each other. So to hit a rabbit running alongside the train you must aim straight at it.

If there were no air, that is all simple enough. But every rifleman knows to his cost that though he fixes both himself and his target tightly to the ground, so as to destroy all aberration proper, yet a current of air is very competent to introduce a kind of spurious aberration of its own, which may be called windage; and that he must not aim at the target if he wants to hit it, but must aim a little in the eye of the wind.

So much from the shooter's point of view. Now attend to the point of view of the target.

Consider it made of soft enough material to be completely penetrated by the bullet, leaving a longish hole wherever struck. A person behind the target, whom we may call a marker, by applying his eye to the hole immediately after

INFLUENCE OF MOTION

the hit, may be able to look through it at the shooter, and thereby to spot the successful man. I know that this is not precisely the function of an ordinary marker, but it is more complete than his ordinary function. All he does usually is to signal an impersonal hit; some one else has to record the identity of the shooter. I am rather assuming a volley of shots, and that the marker has to allocate the hits to their respective sources by means of the holes made in the target.

Well, will he do it correctly? Assuming, of course, that he can do so if everything is stationary, and ignoring all curvature of path, whether vertical or horizontal curvature. If you think it over you will perceive that a wind will not prevent his doing it correctly; the line of hole will point to the shooter along the path of his bullet, though it will not point along his line of aim. Also, if the shots are fired from a moving ship, the line of hole in a stationary target will point to the position the gun occupied at the instant the shot was fired, though it may have moved since then. In neither of these cases (moving medium and moving source) will there be any error.

But if the *target* is in motion, on an armoured train for instance, then the marker will be at fault. The hole will not point to the man who fired the shot, but to an individual ahead of him. *The source will appear to be displaced in the*

THE ETHER OF SPACE

direction of the observer's motion. This is common aberration. It is the simplest thing in the world. The easiest illustration of it is that when you run through a vertical shower, you tilt your umbrella forward; or, if you have not got one, the drops hit you in the face; more accurately, your face as you run forward hits the drops. So the shower appears to come from a cloud ahead of you, instead of from one overhead.

We have thus three motions to consider, that of the source, of the receiver, and of the medium; and, of these, only motion of receiver is able to cause an aberrational error in fixing the position of the source.

So far we have attended to the case of projectiles, with the object of leading up to light. But light does not consist of projectiles, it consists of waves; and with waves matters are a little different. Waves crawl through a medium at their own definite pace; they cannot be *flung* forward or sideways by a moving source; they do not move by reason of an initial momentum which they are gradually expending, as shots do; their motion is more analogous to that of a bird or other self-propelling animal, than it is to that of a shot. The motion of a wave in a moving medium may be likened to that of a rowing-boat on a river. It crawls forward with the water, and it drifts with the water;

INFLUENCE OF MOTION

its resultant motion is compounded of the two, but it has nothing to do with the motion of its source. A shot from a passing steamer retains the motion of the steamer as well as that given it by the powder. It is projected, therefore, in a slant direction. But a boat lowered from the side of a passing steamer, and rowing off, retains none of the motion of its source; it is not projected, it is self-propelled. That is like the case of a wave.

The diagram illustrates the difference. Fig. 1 shows a moving cannon or machine-gun, moving with the arrow, and firing a succession of shots which share the motion of the cannon as well as their own, and so travel slant. The shot fired from position 1 has reached A, that fired from position 2 has reached B, and that fired from position 3 has reached C, by the time the fourth shot is fired at D. The line A B C D is a prolongation of the axis of the gun; it is the line of aim, but it is not the line of fire; all the shots are travelling aslant this line, as shown by the arrows. There are thus two directions to be distinguished. There is the row of successive shots, and there is the path of any one shot. These two directions enclose an angle. It may be called an aberration angle, because it is due to the motion of the source, but it need not give rise to any aberration. True direction may still be perceived from the point of view of the receiver.

THE ETHER OF SPACE

To prove this let us attend to what is happening at the target. The first shot is supposed to be entering at A, and if the target is stationary will leave it at Y. A marker looking along Y A will see the position whence the shot was fired. This may be likened to a stationary observer looking at a moving star. He sees it where and as it was when the light started on its long journey. He

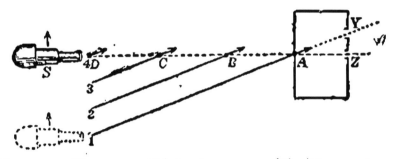

Fig. 1.—Shots or Disturbances with Momentum from a Moving Gun.

does not see its present position, but there is no reason why he should. He does not see its physical state or anything as it is now. He sees it as it was when it sent the information which he has just received. There is no aberration caused by motion of source.

But now let the receiver be moving at same pace as the gun, as when two grappled ships are firing into each other. The motion of the target carries the point Y forward, and the shot A leaves it at Z, because Z is carried to where Y was. So in that case the marker looking along

INFLUENCE OF MOTION

Z A will see the gun, not as it was when firing, but as it is at the present moment; and he will see likewise the row of shots making straight for him. This is like an observer looking at a terrestrial object. Motion of the earth does not disturb ordinary vision.

Fig. 2 shows as nearly the same sort of thing as possible for the case of emitted waves. The tube is a source emitting a succession of disturbances without momentum. A B C D may be thought of as horizontally flying birds, or as crests of waves, or as self-swimming torpedoes; or they may even be thought of as bullets, if the gun stands still every time it fires, and only moves between whiles.

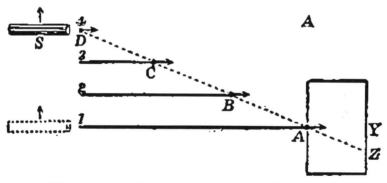

Fig. 2.—Waves or Disturbances without Momentum from a Moving Source.

The line A B C D is now neither the line of fire nor the line of aim: it is simply the locus of disturbances emitted from the successive positions 1 2 3 4.

THE ETHER OF SPACE

A stationary target will be penetrated in the direction A Y, and this line will point out the correct position of the source when the received disturbance started. If the target moves, a disturbance entering at A may leave it at Z, or at any other point according to its rate of motion; the line Z A does not point to the original position of the source, and so there will be aberration when the target moves. Otherwise there would be none.

Now, Fig. 2 also represents a parallel beam of light travelling from a moving source, and entering a telescope or the eye of an observer.

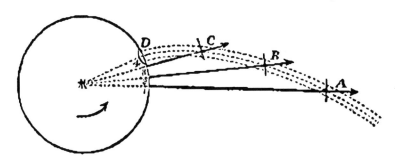

Fig. 3.—Beam from a Revolving Lighthouse.

The beam lies along A B C D, but this is not the direction of vision. The direction of vision, to a stationary observer, is determined not by the locus of successive waves, but by the path of each wave. A ray may be defined as the path of a labelled disturbance. The line of vision is Y A 1, and coincides with the line of aim;

INFLUENCE OF MOTION

which in the projectile case (Fig. 1) it did not.

The case of a revolving lighthouse, emitting long parallel beams of light and brandishing them rapidly round, is rather interesting. Fig. 3 may assist the thinking out of this case. Successive disturbances A, B, C, D, lie along a spiral curve, the spiral of Archimedes; and this is the shape of the beams, as seen illuminating the dust particles, though the pitch of the spiral is too gigantic to be distinguished from a straight line. At first sight it might seem as if an eye looking along those curved beams would see the lighthouse slightly out of its true position; but it is not so. The true rays or actual paths of each disturbance are truly radial; they do not coincide with the apparent beam. An eye looking at the source will not look tangentially along the beam, but will look along A S, and will see the source in its true position. It would be otherwise for the case of projectiles from a revolving turret.

Thus, neither translation of star nor rotation of sun can affect direction. There is no aberration so long as the receiver is stationary.

But what about a wind, or streaming of the medium past source and receiver, both stationary? Look at Fig. 1 again. Suppose a row of stationary cannon firing shots, which get blown by a cross wind along the slant 1 A Y

THE ETHER OF SPACE

(neglecting the curvature of path which would really exist): still the hole in the target fixes the gun's true position, the marker looking along Y A sees the gun which fired the shot. There is no true deviation from the point of view of the receiver, provided the drift is uniform everywhere, although the shots are blown aside and the target is not hit by the particular gun aimed at it.

With a moving cannon combined with an opposing wind, Fig. 1 would become very like Fig. 2.

(N.B.—The actual case, even without complication of spinning, etc., but merely with the curved path caused by steady wind-pressure, is not so simple, and there would really be an aberration or apparent displacement of the source toward the wind's eye: an apparent exaggeration of the effect of wind shown in the diagram.)

In Fig. 2 the result of a wind is much the same, though the details are rather different. The medium is supposed to be drifting downward, across the field. The source may be taken as stationary at S. The horizontal arrows show the direction of waves *in the medium;* the dotted slant line shows their resultant direction. A wave centre drifts from D to 1 in the same time as the disturbance reaches A, travelling down the slant line D A. The angle between dotted and full lines is the angle between ray and wave

INFLUENCE OF MOTION

normal. Now, *if the motion of the medium inside the receiver is the same as it is outside*, the wave will pass straight on along the slant to Z, and the true direction of the source is fixed. But if the medium inside the target or telescope is stationary, the wave will cease to drift as soon as it gets inside—under cover, as it were; it will proceed along the path it has been really pursuing *in the medium* all the time, and make its exit at Y. In this latter case—of different motion of the medium inside and outside the telescope—the apparent direction, such as Y A, is not the true direction of the source. *The ray is in fact bent where it enters the differently moving medium* (as shown in Fig. 4).

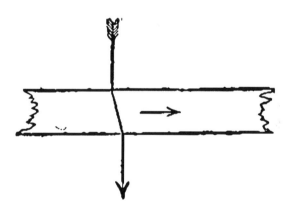

Fig. 4.—Ray through a Moving Stratum.

A slower moving stratum bends an oblique ray, slanting with the motion, in the same direction as if it were a denser medium. A quicker stratum bends it oppositely. If a medium is both denser and quicker moving, it is possible for the two bendings to be equal and opposite, and thus for a ray to go on straight. Parenthetically, I may say that this is precisely what happens, on Fresnel's theory, down the axis of a water-filled telescope exposed to the general terrestrial ether drift.

In a moving medium waves do not advance in their normal direction, they advance slantways. The direction of their advance is properly called a ray. The ray does not coincide with the wave-normal in a moving medium.

All this is well shown in Fig. 5.

S is a stationary source emitting successive waves, which drift as spheres to the right. The wave which has reached M has its centre at C, and C M is its normal; but the disturbance, M, has really travelled along S M, which is therefore the ray. It has advanced as a wave from S to P, and has drifted from P to M. Disturbances subsequently emitted are found along the ray, precisely as in Fig. 2. A stationary telescope receiving the light will point straight at S. A mirror, M, intended to reflect the light straight back must be set normal to the ray, not tangential to the wave front.

The diagram also equally represents the case of a moving source in a stationary medium. The source, starting at C, has moved to S, emitting waves as it went; which waves, as emitted, spread

INFLUENCE OF MOTION

out as simple spheres from the then position of source as centre. Wave-normal and ray now coincide: S M is not a ray, but only the locus of successive disturbances. A stationary telescope

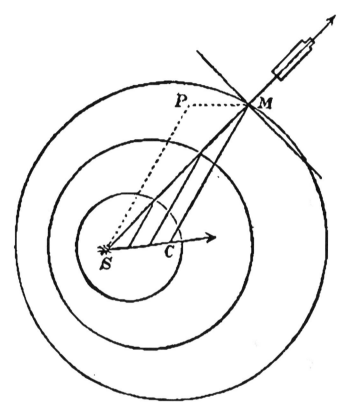

FIG. 5.—Successive Wave Fronts in a Moving Medium.

would look not at S, but along M C to a point where the source was when it emitted the wave M; a moving telescope, if moving at same rate as source, will look at S. Hence S M is sometimes called the *apparent* ray. The angle S M C is the

THE ETHER OF SPACE

aberration angle, which in Chap. X we denote by ε.

Fig. 6 shows normal reflection for the case of a moving medium. The mirror M reflects light received from S_1 to a point S_2 just in time to catch the source there if that is moving with the medium.

Parenthetically, I may say that the time taken on the double journey, $S_1 M S_2$, when the medium is moving, is not quite the same as the double journey S M S, when all is stationary; and that this is the principle of Michelson's great experiment; which must be referred to later.

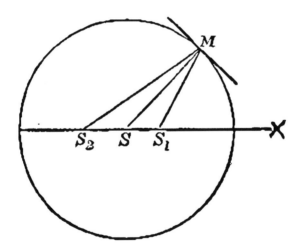

Fig. 6.—Normal Reflection in Moving Medium.
The angle M S X is the angle θ in the theory of Michelson's experiment described in Chapter IV.

The ether stream we speak of is always to be considered merely as one relative to matter. Absolute velocity of matter means velocity through

INFLUENCE OF MOTION

the ether—which is stationary. If there were no such physical standard of rest as the ether—if all motion were relative to matter alone—then the contention of Copernicus and Galileo would have had no real meaning.

IV

EXPERIMENTS ON THE ETHER

WE have arrived at this: that a uniform ether stream all through space causes no aberration, no error in fixing direction. It blows the waves along, but it does not disturb the line of vision.

Stellar aberration exists, but it depends on motion of observer, and on motion of observer only. Etherial motion has no effect upon it; and when the observer is stationary with respect to object, as he is when using a terrestrial telescope, there is no aberration at all.

Surveying operations are not rendered the least inaccurate by the existence of a universal etherial drift; and they therefore afford no means of detecting it.

But observe that everything depends on the ether's motion being uniform everywhere, inside as well as outside the telescope, and along the whole path of the ray. If stationary anywhere it must be stationary altogether: there must be no boundary between stationary and moving

EXPERIMENTS ON THE ETHER

ether, no plane of slip, no quicker motion even in some regions than in others. For (referring back to the remarks preceding Fig. 4) if the ether in receiver is stagnant while outside it is moving, a wave which has advanced and drifted as far as the telescope will cease to drift as soon as it gets inside, but will advance simply along the wave normal. And in general, at the boundary of any such change of motion a ray will be bent, and an observer looking along the ray will see the source not in its true position, not even in the apparent position appropriate to his own motion, but lagging behind that position.

Such an aberration as this, a lag or negative aberration, has never yet been observed; but if there is any slip between layers of ether, if the earth carries any ether with it, or if the ether, being in motion at all, is not equally in motion everywhere throughout every transparent substance, then such a lag or negative aberration must occur: in precise proportion to the amount of the carriage of ether by moving bodies (*cf.* p. 63).

On the other hand, if the ether behaves as a perfectly frictionless inviscid fluid, or if for any other reason there is no rub between it and moving matter, so that the earth carries no ether with it at all, then all rays will be straight, aberration will have its simple and well-known

THE ETHER OF SPACE

value, and we shall be living in a virtual ether stream of 19 miles a second, by reason of the orbital motion of the earth.

It may be difficult to imagine that a great mass like the earth can rush at this tremendous pace through a medium without disturbing it. It is not possible for an ordinary sphere in an ordinary fluid. At the surface of such a sphere there is a viscous drag, and a spinning motion diffuses out thence through the fluid, so that the energy of the moving body is gradually dissipated. The persistence of terrestrial and planetary motions shows that etherial viscosity, if existent, is small; or at least that the amount of energy thus got rid of is a very small fraction of the whole. But there is nothing to show that an appreciable layer of ether may not adhere to the earth and travel with it, even though the force acting on it be but small.

This, then, is the question before us:—

Does the earth drag some ether with it? or does it slip through the ether with perfect freedom? (Never mind the earth's atmosphere; the part it plays is known and not important.)

In other words, is the ether wholly or partially stagnant near the earth, or is it streaming past us with the opposite of the full terrestrial velocity of nineteen miles a second? Surely if we are living in an ether stream of this rapidity we

EXPERIMENTS ON THE ETHER

ought to be able to detect some evidence of its existence.[1]

It is not so easy a thing to detect as you would imagine. We have seen that it produces no deviation or error in direction. Neither does it cause any change of colour or Doppler effect; that is, no shift of lines in spectrum. No steady wind can affect pitch, simply because it cannot blow waves to your ear more quickly than they are emitted. It hurries them along, but it lengthens them in the same proportion, and the result is that they arrive at the proper frequency. The precise effects of motion on pitch are summarised in the following table:—

Changes of Frequency due to Motion

Source approaching shortens waves.
Receiver approaching alters relative velocity.
Medium flowing alters both wave-length and velocity in exactly compensatory manner.

What other phenomena may possibly result from motion? Here is a list:—

Phenomena resulting from Motion

(1) Change or apparent change in direction; observed by telescope, and called aberration.

[1] The word "stationary" is ambiguous. I propose to use "stagnant," as meaning stationary with respect to the earth—*i.e.*, as opposed to stationary in *space*.

THE ETHER OF SPACE

(2) Change or apparent change in frequency; observed by spectroscope, and called Doppler effect.

(3) Change or apparent change in time of journey; observed by lag of phase or shift of interference fringes.

(4) Change or apparent change in intensity; observed by energy received by thermopile.

What we have arrived at so far is the following:—

Motion of either source or receiver can alter frequency; motion of receiver can alter apparent direction; motion of the medium can do neither.

But the question must be asked, can it not hurry a wave so as to make it arrive out of phase with another wave arriving by a different path, and thus produce or modify interference effects?

Or again, may it not carry the waves down stream more plentifully than up stream, and thus act on a pair of thermopiles, arranged fore and aft at equal distances from a source, with unequal intensity?

And once more, perhaps the laws of reflection and refraction in a moving medium are not the same as they are if it be at rest. Then, moreover, there is double refraction, colours of thin plates and thick plates, polarisation angle, rotation of the plane of polarisation; all sorts of optical phenomena that need consideration.

EXPERIMENTS ON THE ETHER

It may have to be admitted, perhaps, that in empty space the effect of an ether drift is difficult to detect, but will not the presence of dense matter—especially the passage through dense transparent matter—make the detection easier? So a great number of questions arise, all of which have been, from time to time, seriously discussed.

Interference.

As an instance of such discussion, consider No. 3 of the phenomena tabulated above. I expect that every reader understands interference, but I may just briefly say that two similar sets of waves "interfere" whenever and wherever the crests of one set coincide with and obliterate the troughs of the other set. Light advances in any given direction when crests in that direction are able to remain crests, and troughs to remain troughs. But if we contrive to split a beam of light into two halves, to send them round by different paths, and make them meet again, there is no guarantee that crest will meet crest and trough trough; it may be just the other way in some places, and wherever that opposition of phase occurs there there will be local obliteration or "interference." Two reunited half-beams of light may thus produce local stripes of darkness, and these stripes are called interference bands.

THE ETHER OF SPACE

It is not to be supposed that there is any *destruction* of light, or any dissipation of energy: it is merely a case of redistribution.

The bright parts are brighter just in proportion as the dark parts are darker. The screen is illuminated in stripes and no longer uniformly, but its total illumination is the same as if there were no interference.

Projection of Interference Bands.

It is not easy to project these interference bands on a screen so as to make them visible to an audience, partly because the bands or stripes of darkness are exceedingly narrow; indeed, I had not previously seen the experiment attempted. But by means of what I call an interference kaleidoscope, consisting of two mirrors set at an angle with a third semi-transparent mirror between them, it is possible to get the bands remarkably clear and bright, so that they can readily be projected: and I showed these at a lecture to the Royal Institution of Great Britain in 1892.

Each mirror is mounted on a tripod with adjustable screw feet, which stand on a thick iron slab, which again rests on hollow india-rubber balls. Looking down on the mirrors the plan is as in the diagram Fig. 7, which indicates sufficiently the geometry of the arrangement,

EXPERIMENTS ON THE ETHER

and shows that the two half beams, into which the semi-transparent plate divides the light, will each travel round the same contour A B C in opposite directions, and will then reunite and travel together toward the point of the arrow.

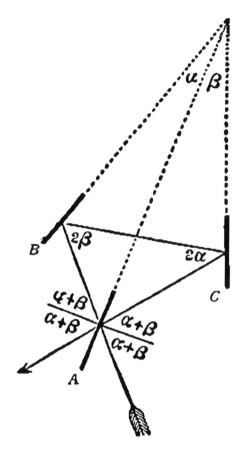

FIG. 7.—Plan of Interference Kaleidoscope with three mirrors.

The arrow-feather ray is bifurcated at A by a semi-transparent mirror of thinly silvered glass; and the two halves reunite along the arrow-head after traversing a triangular contour A B C in opposite directions. The simple geometrical relations which permit this are sufficiently indicated in the figure. The arrangement would suit Fizeau's experiment.

53

THE ETHER OF SPACE

A parallel beam from an electric lantern, when thus treated, depicts bright and broad interference bands on a screen. And the arrangement is very little sensitive to disturbance, because the paths of the two halves of the beam are identical, and because of the mounting. A piece of good glass can be interposed without disturbance, and the table can be struck a heavy blow without confusing the bands.

The only regular and orderly way of causing a shift of the bands is to accelerate one half of the beam and to retard the other half by moving a transparent substance along the contour. For instance, let the sides of the triangle A B C, or one of them, consist of a tube of water in which a rapid stream is maintained; then the stream has a chance of accelerating one half the beam and retarding the other half, thereby shifting the fringes from their normal position by a measurable amount. This is the experiment made in 1859 by Fizeau. (Appendix 3.)

Now that most interesting and important, and I think now well-known, experiment of Fizeau proves quite simply and definitely that if light be sent along a stream of water, travelling inside the water as a transparent medium, it will go quicker with the current than against it.

You may say that is only natural; a wind assists sound one way and retards it the opposite way. Yes, but then sound travels in air; and

EXPERIMENTS ON THE ETHER

spect to the earth's motion; but shift was seen.

Interference methods all fail t trace of relative motion between pa

Try other phenomena, then. T The index of refraction of glass is k pend on the ratio of the speed of ligh the speed inside the glass. If, then, t streaming through glass, the velocity will be different inside according as with the stream or against it, and so the refraction may be different. Arago was the first to try this experiment by placing an achromatic

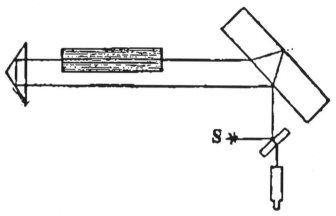

FIG. 9.—Arrangement of Mascart and Jamin.

A modification of Fig. 8, with the beam split definitely into two halves by reflection from a thick glass plate and reunited before observation. The two half beams go through stagnant water in opposite directions.

prism in front of a telescope on a mural circle and observing the deviation it produced on stars.

Observe that it was an *achromatic* prism,

treating all wave-lengths alike; he looked at the *deviated* image of a star, not at its *dispersed* image or spectrum—else he might have detected the change-of-frequency-effect due to motion of source or receiver first actually seen by Sir W. Huggins. I do not think Arago would have seen it, because I do not suppose his arrangements were delicate enough for that very small effect; but there is no error in the conception of his experiment, as Professor Mascart has inadvertently suggested there was.

Then Maxwell repeated the attempt in a much more powerful manner, a method which could have detected a very minute effect indeed, and Mascart has also repeated it in a simple form. All are absolutely negative.

Well, then, what about aberration? If one looks through a moving stratum, say a spinning glass disk, there ought to be a shift caused by the motion (*see* Fig. 4). That particular experiment has not been tried, but I entertain no doubt about its result, though a high speed and considerable thickness of glass or other medium would be necessary to produce even a microscopic apparent displacement of objects seen through it.

But the speed of the earth is available, and the whole length of a telescope tube may be filled with water; surely that is enough to displace rays of light appreciably.

Sir George Airy tried it at Greenwich on a star,

EXPERIMENTS ON THE ETHER

with an appropriate zenith-sector full of water. Stars were seen through the water-telescope precisely as through an air telescope. A negative result again! (The theory is fully dealt with in Chapter X and Appendix 3.)

Stellar observations, however, are unnecessarily difficult. Fresnel had pointed out that a terrestrial source of light would do just as well. He had also (being a man of exceeding genius) predicted that nothing would happen. Hoek has now tried it in a perfect manner and nothing did happen.

But these facts are not at all disconcerting; they are just what ought to be anticipated, in the light of true theory. The absence of all effect caused by stagnant dense matter inserted in the path of a beam of light, that is of dense transparent matter not artificially moved with reference to the earth—or rather with reference to source and receiver—is explicable on Fresnel's theory concerning the behaviour of ether inside matter.

If the index of refraction of the matter is called μ, that means that the speed of light inside it is $\frac{1}{\mu}$th of the speed outside or in vacuo. And that is only another way of saying that the virtual etherial density inside it is represented by μ^2, since the velocity of waves is inversely as the square root of the density of the medium which conveys them; the elasticity being reckoned as constant, and the same inside as out.

THE ETHER OF SPACE

But then if the ether is incompressible its density must really be constant, so how can it be denser inside matter than it is outside? The answer is that presumably the ether is not really extra dense, but is, as it were, *loaded* by the matter. The atoms of matter, or the constituent electrons, must be presumed to be shaken by the passage of the waves of light, as they obviously are in fluorescent substances; and accordingly the speed of propagation will be lessened by the extra loading which the waves encounter. It is not a real increase of density, but a virtual increase, which is really due to the addition of a certain fraction of material inertia to the inertia of the ether itself. The density of ether outside being 1, and that of the loaded ether inside being μ^2, the effect of the load is expressible as $\mu^2 - 1$, while the free ether is the same inside as out.

Suppose now that the matter is moved along. The extra loading, being part of the matter, of course travels with it, and thereby affects the speed of light to the extent of the load—that is to say, by an amount proportional to $\mu^2 - 1$ as contrasted with μ^2.

This is Fresnel's predicted ratio $(\mu^2 - 1) : \mu^2$, or $1 - \frac{1}{\mu^2}$; and in Fizeau's experiment with running water—especially as repeated later, with modern accuracy, by Michelson—this represents exactly the amount of observed effect upon the light.

EXPERIMENTS ON THE ETHER

But if, instead of running water, stagnant water is used—that is stationary with respect to the earth, though still moving violently through the ether—then the (μ^2-1) effect of the load will be fixed to the matter, and can produce no extra or motile effect. The only part that could produce an effect of that kind would be the free ether, of density 1. But then this—on the above view—is absolutely stationary, not being carried along by the earth at all; hence this can give no effect either. Consequently the whole effect of an ether-drift past the earth is zero, on optical experiments, according to the theory of Fresnel; and that is exactly what all the experiments just described have confirmed.

Since then Professor Mascart, with great pertinacity, has attacked the phenomena of thick plates, Newton's rings, double refraction, and the rotatory phenomenon of quartz; but he has found absolutely nothing attributable to a stream of ether past the earth.

The only positive result ever supposed to be attained was in a very difficult polarisation observation by Fizeau in 1859. Unless this has been repeated, it is safest to ignore it; but I believe that Lord Rayleigh has repeated it, and obtained a negative result.

Fizeau also suggested, but did not attempt, what seems an easier experiment, with fore and aft thermopiles and a source between them, to observe the drift of a medium by its convection of energy; but arguments based on the law of exchanges[1] tend to show, and do show as I think, that a probable alteration of radiating power due to motion through a medium would just compensate the effect otherwise to be expected.

We may summarise most of these statements as follows:—

Summary.

Source alone moving produces . . .
- A real and apparent change of wave-length.
- A real but not apparent error in direction.
- No lag of phase or change of intensity, except that appropriate to altered wave-length.

Medium alone moving, or source and receiver moving together, produces . . .
- No change of frequency.
- No error in direction.
- A real lag of phase, but undetectable without control over the medium.
- A change of intensity corresponding to different distance, but compensated by change of radiating power.

[1] Lord Rayleigh, "Nature," March 25, 1892.

EXPERIMENTS ON THE ETHER

Receiver alone moving produces . . . $\begin{cases} \text{An apparent change of wavelength.} \\ \text{An apparent error in direction.} \\ \text{No change of phase or of intensity, except that appropriate to different virtual velocity of light.} \end{cases}$

I may say, then, that not a single optical phenomenon is able to show the existence of an ether stream near the earth. All optics go on precisely as if the ether were stagnant with respect to the earth.

Well, then, perhaps it *is* stagnant. The experiments I have quoted do not prove that it is so. They are equally consistent with its perfect freedom and with its absolute stagnation, though they are not consistent with any intermediate position. Certainly, if the ether were stagnant nothing could be simpler than their explanation.

The only phenomena then difficult to explain would be those depending on light coming from distant regions through all the layers of more or less dragged ether. The theory of astronomical aberration would be seriously complicated; in its present form it would be upset (p. 47). But it is never wise to control facts by a theory; it is better to invent some experiment that will give a

THE ETHER OF SPACE

different result in stagnant and in free ether. None of those experiments so far described are really discriminative. They are, as I say, consistent with either hypothesis, though not very obviously so.

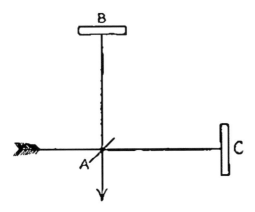

FIG. 10.—The course of the light and of the two half beams in Michelson's most famous experiment.

The light is split at A, one half sent toward B and back, the other half to C and back. (Compare with Fig. 7.)

Michelson Experiment.

Mr. Michelson, however, of the United States, invented a plan that looked as if it really would discriminate; and, after overcoming many difficulties, he carried it out. It is described in the *Philosophical Magazine* for 1887.

Michelson's famous experiment consists in looking for interference between two half beams of light, of which one has been sent to and fro *across* the line of ether drift, and the other has been sent to and fro *along* the line of ether drift.

EXPERIMENTS ON THE ETHER

A semi-transparent mirror set at 45° is employed to split the beam, and a pair of normal and ordinary mirrors, set perpendicular to the two half beams, are employed to return them back whence they came, so that they can enter the eye through an observing telescope.

It differs essentially from the interference kaleidoscope, Fig. 7, inasmuch as there is now no luminous path B C, and no contour enclosed by the light. Each half beam goes to and fro on its own path, and these paths, instead of being coincident, are widely separate—one north and south, for instance, and the other east and west.

Under these conditions the bands are much more tremulous than they were in the arrangement of Fig. 7, and are subject to every kind of disturbance. The apparatus has to be excessively steady, and no fluctuation even of temperature must be permitted in the path of either beam. To secure this, the source, the mirrors, and the observing telescope were all mounted upon a massive stone slab; and this was floated in a bath of mercury.

The slab could then be slowly turned round, so that sometimes the path A B and sometimes the path A C lay approximately along or athwart the direction of the earth's motion in space.

And inasmuch as the motion along would take

THE ETHER OF SPACE

a little longer than the motion across, though everything else was accurately the same, some shift of the interference bands might be expected as the slab rotated.

But whereas in all the experiments previously described the effect looked for was a first-order effect, of magnitude one in ten or twenty thousand—depending, that is to say, on the first power of the ratio of speed of earth to speed of light—the effect now to be expected depends on the *square* of that same ratio, and therefore cannot be greater, even in the most favourable circumstances, than 1 part in a hundred million.

It is easy to realise, therefore, that it is an exceptionally difficult experiment, and that it required both skill and pertinacity to perform it successfully.

That it is an exceptionally difficult experiment will be realised when I say that it would fail in conclusiveness unless one part in 400 millions could be clearly detected.

Mr. Michelson reckons that by his latest arrangement he could see 1 in 4000 millions if it existed (which is equivalent to detecting an error of $\frac{1}{1000}$th of an inch in a length of 60 miles); but he saw nothing. Everything behaved precisely as if the ether was stagnant; as if the earth carried with it all the ether in its immediate neighbourhood. And that was his conclusion.

EXPERIMENTS ON THE ETHER

Theory of Michelson Experiment.

The theory of the Michelson experiment can be expressed thus: its optical diagram being the same as is expressed geometrically in Fig. 6.

If a relatively fixed source and receiver move through the ether with velocity u, such that $u/v = a$ the aberration constant; then the time of any to-and-fro journey S M, inclined at angle θ to the direction of the drift, is increased, above what it would be if there were no drift, in the ratio

$$\frac{\sqrt{(1 - a^2 \sin^2\theta)}}{1 - a^2}$$

This follows from merely geometrical considerations.

Hence if a ray is split, and half sent so that $\theta = o$ while the other half is sent so that $\theta = 90$ (as in Fig. 10), the one will lag behind the other by a distance $\frac{1}{2}a^2$ times the distance travelled; which, though very small, may be a perceptible fraction of a wave-length, and therefore may cause a perceptible shift of the bands.

But when the experiment is properly performed, no such shift is observed.

The experiment thus seems to prove that there is no motion through the ether at all, that there is no etherial drift past the earth, that the ether immediately in contact with the earth is

THE ETHER OF SPACE

stagnant—or that the earth to that extent carries all neighbouring ether with it.

If we wish to evade this conclusion, there is no easy way of doing so. For it depends on no doubtful properties of transparent substances, but on the straightforward fundamental principle underlying all such simple facts as that— It takes longer to row a certain distance and back, up and down stream, than it does to row the same distance in still water; or that it takes longer to run up and down a hill than to run the same distance laid out flat; or that it costs more to buy a certain number of oranges at three a penny and an equal number at two a penny than it does to buy the whole lot at five for twopence.

Hence, although there may be *some* way of getting round Mr. Michelson's experiment, there is no obvious way; and if the true conclusion be not that the ether near the earth is stagnant, it must lead to some other important and unknown fact.

That fact has now come clearly to light. It was first suggested by the late Prof. G. F. Fitz-Gerald, of Trinity College, Dublin, while sitting in my study at Liverpool and discussing the matter with me. The suggestion bore the impress of truth from the first. It independently occurred also to Prof. H. A. Lorentz, of Leiden, into whose theory it completely fits, and who has

EXPERIMENTS ON THE ETHER

brilliantly worked it into his system. It may be explained briefly thus:—

Electric charges in motion constitute an electric current. Similar charges repel each other, but currents in the same direction attract. Consequently two similar charges moving in parallel lines will repel each other less than if stationary—less also than if moving one after the other in the same line. Likewise two opposite charges, a fixed distance apart, attract each other less when moving side by side than when chasing each other. The modification of the static force, thus caused, depends on the squared ratio of their joint speed to the velocity of light.

Atoms of matter are charged; and cohesion is a residual electric attraction (*see* end of Appendix 1). So when a block of matter is moving through the ether of space its cohesive forces across the line of motion are diminished, and consequently in that direction it expands, by an amount proportioned to the square of aberration magnitude.

A light journey, to and fro, across the path of a relatively moving medium is slightly quicker than the same journey, to and fro, along (*see* p. 67). But if the journeys are planned or set out on a block of matter, they do not remain quite the same when it is conveyed through space: the journey across the direction of motion becomes longer than the other journey, as we have just seen. And the extra distance compensates or neutralises the extra speed; so that light takes the same time for both.

V

SPECIAL EXPERIMENT ON ETHERIAL VISCOSITY

THE balance of evidence at this stage seems to incline in the sense that there is no ether drift, that the ether near the earth is stagnant, that the earth carries all or the greater part of the neighbouring ether with it—a view which, if true, must singularly complicate the theory of ordinary astronomical aberration: as is explained at the beginning of the last chapter.

But now put the question another way. *Can* matter carry neighbouring ether with it when it moves? Abandon the earth altogether; its motion is very quick but too uncontrollable, and it always gives negative results. Take a lump of matter that you can deal with, and see if it pulls any ether along.

That is the experiment which I set myself to perform, and which in the course of the years 1891–97 I performed. It may be thus described in essence:—

Take a steel disk, or rather a couple of large steel disks a yard in diameter clamped together

SPECIAL EXPERIMENT

with a space between. Mount the system on a vertical axis, and spin it like a teetotum as fast as it will stand without flying to pieces. Then take a parallel beam of light, split it into two by a semi-transparent mirror, M, a piece of glass silvered so thinly that it lets half the light through and reflects the other half, somewhat as in Fig. 7; and send the two halves of this split beam round and round in opposite directions in the space between the disks. They may thus travel a distance of 20 or 30 or 40 feet. Ultimately they are allowed to meet and enter a telescope. If they have gone quite identical distances they need not interfere, but usually the distances will differ by a hundred-thousandth of an inch or so, which is quite enough to bring about interference.

The mirrors which reflect the light round and round between the disks are shown in Fig. 11. If they form an accurate square the last two images will coincide, but if the mirrors are the least inclined to one another at any unaliquot part of 360° the last image splits into two, as in the kaleidoscope is well known, and the interference bands may be regarded as resulting from those two sources. The central white band bisects normally the distance between them, and their amount of separation determines the width of the bands. There are many interesting optical details here, but I shall not go into them.

THE ETHER OF SPACE

The thing to observe is whether the motion of the disks is able to replace a bright band by a dark one, or vice versa. If it does, it means

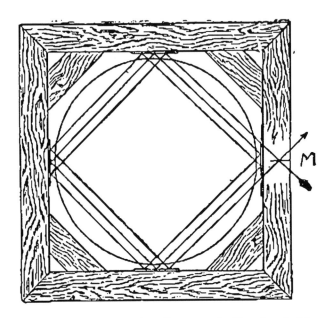

FIG. 11.—Diagrammatic Plan of Optical Frame for Ether Machine; with Steel Disks, one yard in diameter, inside the frame. (The actual apparatus is shown in Figs. 13 and 14 and Fig. 12.)

M is a semi-transparent mirror, reflecting half an incident beam and transmitting the other half. The two half beams each go three times round the square contour, in opposite directions, and then reunite. It is an extension of the idea of Fig. 7.

that one of the half beams—*viz.*, that which is travelling in the same direction as the disks—is helped on a trifle, equivalent to a shortening of journey by some quarter millionth of an inch or so in the whole length of 30 feet; while the other half beam—*viz.*, that travelling against the

SPECIAL EXPERIMENT

motion of the disks—is retarded, or its path virtually lengthened, by the same amount.

If this acceleration and retardation actually occur, waves which did not interfere on meeting before the disks moved, will interfere now; for one will arrive at the common goal half a length behind the other.

Now a gradual change of bright space to dark, and vice versa, shows itself, to an observer looking at the bands, as a gradual change of position of the bright stripes, or a shift of the bands. A shift of the bands, and especially of the middle white band, which is much more stable than the others, is what we look for. The middle band is, or should be, free from the "concertina"-like motion which is liable to infect the others.

At first I saw plenty of shift. In the first experiment the bands sailed across the field as the disks got up speed until the crosswire had traversed a band and a half. The conditions were such that had the ether whirled at the full speed of the disks I should have seen a shift of three bands. It looked very much as if the light was helped along at half the speed of the moving matter, just as it is inside water.

On stopping the disks the bands returned to their old position. On starting them again in the opposite direction, the bands ought to have shifted the other way too, if the effect was

THE ETHER OF SPACE

genuine; but they did not; they went the same way as before.

The shift was therefore wholly spurious; it was caused by the centrifugal force of the blast of air thrown off from the moving disks. The mirrors and frame had to be protected from this. Many other small changes had to be made, and gradually the spurious shifts have been reduced and reduced, largely by the skill and patience of my assistant, Mr. Benjamin Davies, until presently there was barely a trace of them.

But the experiment is not an easy one. Not only does the blast exert pressure, but at high speeds the churning of the air makes it quite hot. Moreover, the tremor of the whirling machine, in which from four to nine horsepower is sometimes being expended, is but too liable to communicate itself to the optical part of the apparatus. Of course elaborate precautions are taken against this. Although the two parts, the mechanical and the optical, are so close together, their supports are entirely independent. But they have to rest on the same earth, and hence communicated tremors are not absent. They are the cause of most of the slight residual trouble.

The whole experiment is described in fairly full detail in the *Philosophical Transactions of the Royal Society* for 1893 and 1897. And there also are described some further modifications where-

SPECIAL EXPERIMENT

by the whirling disks are electrified—likewise without optical effect, and are also magnetised; or rather a great iron mass, strongly magnetised by a current, is used to replace the steel disks.

The effect was always zero, however, when spurious results were eliminated; and it is clear that at no practicable speed does either electrification or magnetisation confer upon matter any appreciable viscous grip upon the ether. Atoms *must* be able to throw it into vibration, if they are oscillating or revolving at sufficient speed; otherwise they would not emit light or any kind of radiation; but in no case do they appear to drag it along, or to meet with resistance in any uniform motion through it. Only their acceleration is effectual.

In the light of Larmor's electron theory, we know now that acceleration of atoms, or rather of a charge upon an atom, necessarily generates radiation, proportional in amount to the *square* of the acceleration—whether that be tangential or normal. There is no theoretical reason for assuming any influence on uniform velocity. And even the influence on acceleration is exceedingly small under ordinary circumstances. Only during the violence of collision are ether waves freely excited. The present experiment, however, has nothing to do with acceleration: it is a test of viscosity. An acceleration term exists in motion through even a perfect fluid.

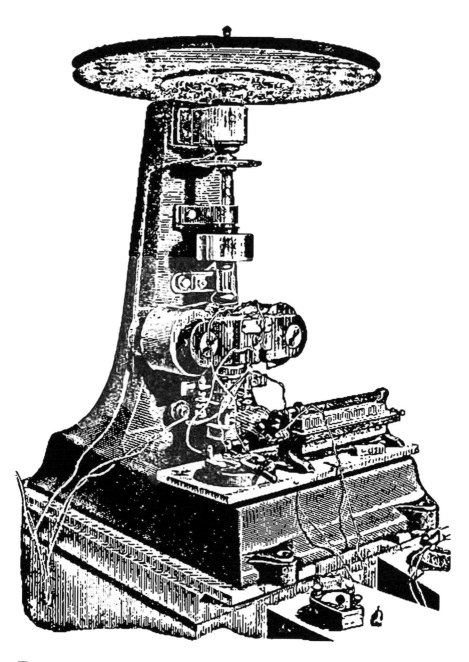

FIG. 12.—General view of whirling part of Ether machine, with pair of steel disks, and motor.

SPECIAL EXPERIMENT

The conclusion at which I arrived in 1892 and 1893 is thus expressed (p. 777 of vol. 184 *Philosophical Transactions of the Royal Society*): "I feel confident either that the ether between the disks is quite unaffected by their motion, or, if affected at all, by something less than the thousandth part. At the same time, so far as rigorous proof is concerned, I should prefer to assert that *the velocity of light between two steel plates moving together in their own plane an inch apart is not increased or diminished by so much as the $\frac{1}{200}$th part of their velocity.*

That was the conclusion in 1893; but since then observations have been continued, and it is now quite safe to change the $\frac{1}{200}$th into $\frac{1}{1000}$th. The spin was sometimes continued for three hours to see if an effect developed with time; and many other precautions were taken, as briefly narrated in the *Philosophical Transactions* for 1897.

The following illustrations give an idea of the apparatus employed:

Fig. 12 shows a photograph of the whirling machine before being bolted down to its stone pier; with the pair of disks at top ready to be whirled by an armature on the shaft, which is supplied with a current sometimes of nine horsepower. The armature winding was of low resistance, and was specially braced, so as to give high speed without flying out, and without

THE ETHER OF SPACE

generating too much back-E M F. The ampere-meter and volt-meter and the carbon rheostat (in armature circuit), for regulating the speed, are plainly seen. The smooth pulley on the shaft is for applying a brake. The small disk above it is perforated to act as a syren for estimation of speed; but other arrangements for this purpose were subsequently added. The two large disks at top were of the best circular-saw steel; they are somewhat thicker at middle than at edge, and are strongly bolted up between iron cheeks, which are attached to the shaft. The lower end of the shaft is a step-bearing of hardened steel in a vessel of oil. The upper collar is elastic, so as to allow for a steadying teetotum action at high speeds.

Fig. 13 is a photograph of the optical square, which was ultimately to be placed in position surrounding the disks. The slit and collimator are shown; the micrometer end of the observing telescope is out of the picture.

The mirrors on the sides of the square are accurately plane; they are adjustable on geometric principles, and are pressed against their bearings by strong spiral springs. They were made by Hilger.

A drawing of the arrangement is given in Fig. 14, and here the double micrometer eyepiece is visible.

In Fig. 15 the whole apparatus is shown

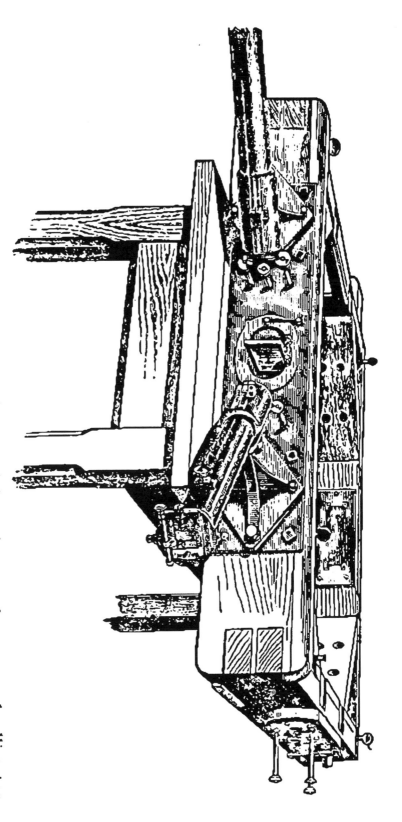

Fig. 13.—General view of optical framework—sustaining mirrors, telescope, and collimator—to surround the disks of the Ether machine. (Compare Fig. 11.)

THE ETHER OF SPACE

mounted. The whirling machine strongly bolted down to a stone pier independent of the floor; the optical frame independently supported by a gallows frame from other piers. The centrifugal mercury speed-indicator is visible in front, and Mr. Davies is regulating the speed. At the back is seen a boiler-plate screen for the observer with his eye at the telescope. (*See* Frontispiece.)

The expense of the apparatus was borne by my friend, the late George Holt, shipowner, of Liverpool.

Fig. 16 exhibits something like the appearance seen in the eye-piece, with the interference bands on each side of the middle band, and with the

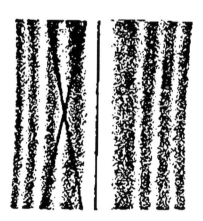

Fig. 16.—Approximate appearance of the interference bands and micrometer Tires as seen in the eye-piece of the telescope of the Ether machine.

micrometer wires set in position—each moved by an independent micrometer head. The straight vertical wire was usually set in the centre of the

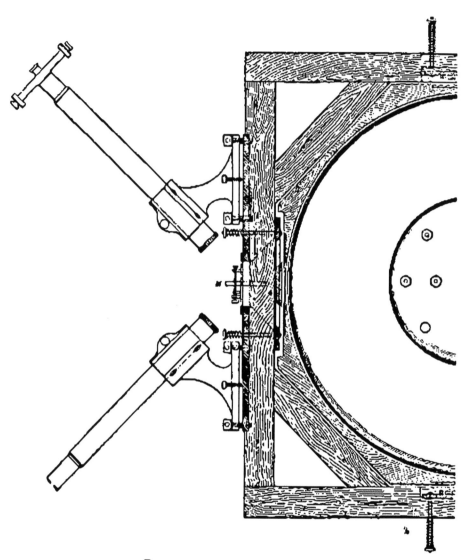

FIG 14 Plan of optical frame with steel disk or parts of frame. O represents one of the panes of optic shewn, and part of the fixing of the rear mirror frame, each mirror held by a brass plate and it is pressed by the spring-bolts shewn

Phil. Trans. 1893.A. *Plate* 31.

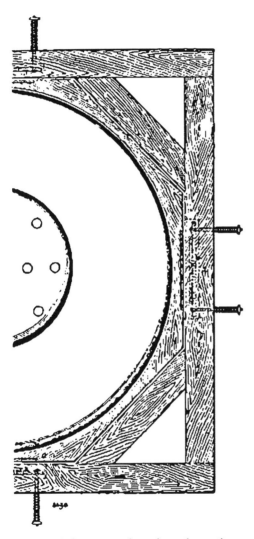

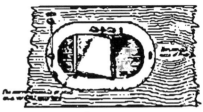

Mode of mounting the semitransparent
mirror M so as to give altitude and
azimuth movement to the reflected beam.

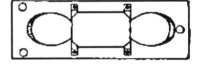

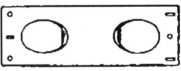

Details of brass plate supporting tooth mirror,
front, side, and back views.
Back view shows the three slots in which
the ends of the supporting screws rest giving a
fine adjustment, the plate being supported by
three rigid pushes and three elastic pulls.

SPECIAL EXPERIMENT

middle white band, and the X wire on the yellow of the first coloured band on one side or the other.

The method of observation now consists in setting a wire of the micrometer accurately in the centre of the middle band, while another wire is usually set on the first band to the left. Then the micrometer heads are read, and the setting repeated once or twice to see how closely and dependably they can be set in the same position. Then we begin to spin the disks, and when they are going at some high speed, measured by a siren note and in other ways, the micrometer wires are reset and read—reset several times and read each time. Then the disks are stopped and more readings are taken. Then their motion is reversed, the wires set and read again; and finally the motion is once more stopped and another set of readings taken. By this means the absolute shift of middle band, and its relative interpretation in terms of wavelength, are simultaneously obtained; for the distance from the one wire to the other, which is often two revolutions of a micrometer head, represents a whole wave-length shift.

In the best experiments I do still often see something like a fiftieth of a band shift; but it is caused by residual spurious causes, for it repeats itself with sufficient accuracy in the same direction when the disks are spun the other way round.

Of real reversible shift, due to motion of the

THE ETHER OF SPACE

ether, I see nothing. I do not believe the ether moves. It does not move at a five-hundredth part of the speed of the steel disks. Further experience confirms and strengthens this estimate, and my conclusion is that such things as circular-saws, flywheels, railway trains, and all ordinary masses of matter do not appreciably carry the ether with them. Their motion does not seem to disturb it in the least.

The presumption is that the same is true for the earth; but the earth is a big body—it is conceivable that so great a mass may be able to act when a small mass would fail. I would not like to be too sure about the earth—at least, not on a strictly experimental basis. What I do feel sure of is that if moving matter disturbs ether in its neighbourhood at all, it does so by some minute action, comparable in amount perhaps to gravitation, and possibly by means of the same property as that to which gravitation is due—not by anything that can fairly be likened to etherial viscosity. So far as experiment has gone, our conclusion is that the viscosity or fluid friction of the ether is zero. And that is an entirely reasonable conclusion.

Magnetisation.

For testing the effect of magnetism, an oblate spheroid was made of specially selected soft iron,

SPECIAL EXPERIMENT

3 feet in diameter, weighing nearly a ton. Its section is shown in Fig. 17. It had an annular channel or groove, half an inch wide and 1 foot deep, round the bottom of which was wound a kilometer of insulated wire to a depth of $4\frac{1}{2}$ inches; the terminals of which were brought out to sliding contacts on the shaft, so that, the whole could be very highly magnetised while it was spinning. Everything was arranged so as to be symmetrical about the central axis.

To the coil of wire, whose resistance was 30 ohms, 110 volts was ordinarily, and 220 volts exceptionally, applied. The magnetic field with 110 volts was about 1800 c.g.s., on the average, all over the main region through which the beam of light circulated.

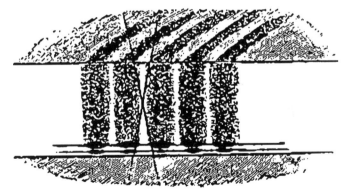

FIG. 18.—Appearance of the interference bands in the channel of the iron spheroid. They were reflected in the upper iron as shown.

This light-bearing space, or gap in the magnetic circuit, was only half an inch wide; and ac-

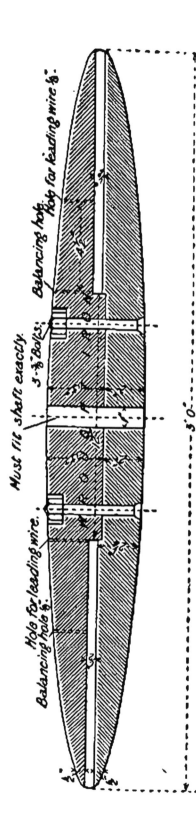

FIG. 17.—Section of oblate spheroid of soft iron for whirling machine, showing arrangement for winding central core with wire so as to be able to magnetise it strongly while spinning inside the optical frame.

SPECIAL EXPERIMENT

cordingly in the eyepiece the iron surfaces could be seen, above and below, as well as the interference bands in the luminous gap. The whole appearance is depicted in Fig. 18.

ELECTRIFICATION.

For the electrification experiment, a third and insulated disk was clamped between the two steel disks and kept electrified to sparking tension. The arrangement is shown diagrammatically on a smaller scale in Fig. 19.

FIG. 19.—Arrangement for electrifying a third or middle steel disk to sparking potential while spinning.

The electrification test was exceptionally easy to apply by connecting the insulated charging pin to a Voss machine in action: because when the disks were spinning and the bands in good condition, the electrification could be instantaneously applied, taken off, reversed, or whatever was desired; and the effect of the sudden lowering of potential by sparks passing between the revolving plates could be exactly looked for.

The conclusion of my second *Philosophical Transactions* paper—that of 1897—is that *neither an electric nor a magnetic transverse field confers viscosity upon the ether, nor enables moving matter to grip and move it rotationally.*

Question of a Possible Longitudinal Magnetic Drift.

Later I tried a longitudinal magnetic field also; arranging a series of four large electric bobbins or long coils along the sides of a square inscribed at 45° in the optical square (Figs. 11 and 13), so that the light went along their axes.

The details of this experiment have been only partially recorded, but the salient points are to be found stated in the *Philosophical Magazine* for April, 1907, pages 495–500.

The result was again negative; that is to say, a magnetic field causes no perceptible acceleration in a beam of light sent along the lines of force. The extra velocity that could have been observed would have been $\frac{1}{5}$th of a millimeter per second, or 16 miles per hour, for each c.g.s. unit of field intensity.

Another mode of expressing the result is that the difference of magnetic potential applied, namely, a drop of two million c.g.s. units of magnetic potential, does not hurry light along it by so much as $\frac{1}{60}$th part of a wave-length.

There may be reasons for supposing that some much slower drift or conveyance than this

SPECIAL EXPERIMENT

is really caused in the ether by a magnetic field; but if so, the ether must be regarded as so excessively dense that the amount of such a drift for any practicable magnetic field seems almost hopelessly beyond experimental means of detection.

VI

ETHERIAL DENSITY

THIS leads us to enter upon the question of whether it is possible to determine with any approach to accuracy the actual density or massiveness of the ether of space, compared with those forms of matter to which our senses have made us accustomed.

The arguments on which an estimate may be made of the density or massiveness of the ether as compared with that of matter depend on the following considerations, the validity of which again is dependent upon an electrical theory of matter. In this theory, or working hypothesis, an assumption has to be made: but it is one for which there is a large amount of justification, and the reasons for it are given in many books—among others in my book on *Electrons*, and likewise at the end of the new edition of *Modern Views of Electricity*, also in my *Romanes Lecture*, published by the Clarendon Press in 1903. Put briefly, the assumption is that matter is composed, in some way or other, of electrons; which

ETHER DENSITY

again must be considered to be essentially peculiarities, or singularities, or definite structures, in the ether itself. Indeed, a consideration of electrons alone is sufficient for the argument, provided it be admitted that they have the mass which experiment shows them to possess, and the size which electrical theory deduces for them: the basis of the idea—which, indeed, is now experimentally proved—being that their inertia is due to their self-induction—*i.e.*, to the magnetic field with which they must be surrounded as long as they are in motion.

The mass, or inertia, of an electron is comparable to the thousandth part of that of the atom of hydrogen. Its linear dimension, let us say its diameter, is comparable to the one hundred thousandth part of what is commonly known as molecular or atomic dimensions; which itself is the ten-millionth part of a millimeter.

Hence, the mass and the bulk of an electron being known, its density is determined, provided we can assume that its mass is all dependent on what is contained within its periphery. But that last assumption is one that quite definitely cannot be made: its mass is for the most part outside itself, and has to be calculated by magnetic considerations. (*See* Appendix 2.)

These details are gone into in my paper in the *Philosophical Magazine* for April, 1907, and in Chapter XVII of *Modern Views of Electricity*.

THE ETHER OF SPACE

But without repeating arguments here, it will suffice to say that although the estimates may be made in various ways, differing entirely from each other, yet the resulting differences are only slight; the calculated densities come out all of the same order of magnitude—namely, something comparable to 10^{12} c.g.s. units; that is to say, a million million grammes per cubic centimeter, or, in other words, a thousand tons to the cubic millimeter.

But, throughout, we have seen reason to assert that the ether is incompressible; arguments for this are given in *Modern Views of Electricity*, Chapter I. And, indeed, the fundamental medium filling all space, if there be such, *must*, in my judgment, be ultimately incompressible; otherwise it would be composed of parts, and we should have to seek for something still more fundamental to fill the interstices.

The ether being incompressible, and an electron being supposed composed simply and solely of ether, it follows that it cannot be either a condensation or a rarefaction of that material, but must be some singularity of structure, or some portion otherwise differentiated. It might, for instance, be something analogous to a vortex ring, differentiated kinetically—*i.e.*, by reason of its rotational motion, from the remainder of the ether; or it might be differentiated statically, and be something which would have to be called

ETHER DENSITY

a strain-centre or a region of twist, or something which cannot be very clearly at present imagined with any security; though various suggestions have been made in that direction.

The simplest plan for us is to think of it somewhat as we think of a knot on a piece of string. The knot differs in no respect from the rest of the string, except in its tied-up structure; it is of the same density with the rest, and yet it is differentiated from the rest; and, in order to cease to be a knot, would have to be untied—a process which as yet we have not yet learned how to apply to an electron. If ever such a procedure becomes possible, then electrons will thereby be resolved into the general body of the undifferentiated ether of space—that part which is independent of what we call "matter."

The important notion for present purposes is merely this: that the density of the undifferentiated or simple ether, and the density of the tied-up or beknotted or otherwise modified ether constituting an electron, are one and the same. Hence the argument above given, at least, when properly worked out, tends to establish the etherial density as of the order 10^{12} times that of water.

There ought to be nothing surprising (though I admit that there is something very surprising) in such an estimate; inasmuch as many converging lines of argument tend to show that ordinary

THE ETHER OF SPACE

matter is a very porous or gossamer-like substance, with interspaces great as compared with the spaces actually occupied by the nuclei which constitute it. Our conception of matter, if it is to be composed of electrons, is necessarily rather like the conception of a solar system, or rather of a milky way; where there are innumerable dots here and there, with great interspaces between. So that the average density of the whole of the dots or material particles taken together—that is to say, their aggregate mass compared with the space they occupy—is excessively small.

In the vast extent of the Cosmos, as a whole, the small bulk of actual matter, compared with the volume of empty space, is striking—as we shall show directly; and now on the small scale, among the atoms of matter, we find the conditions to be similar. Even what we call the densest material is of extraordinarily insignificant massiveness as compared with the unmodified ether which occupies by far the greater proportion of its bulk.

When we speak of the density of *matter*, we are really though not consciously expressing the group-density of the modified ether which constitutes matter—not estimated per unit, but per aggregate; just as we might estimate the group or average density of a cloud or mist. Reckoned per unit, a cloud has the density of water; reckoned per aggregate, it is an impalpable filmy structure of hardly any density at all. So it is with a cobweb, so perhaps it is with a comet's tail, so also with the Milky Way, with the cosmos—and, as it now turns out, with ordinary matter itself.

For consider the average density of the material cosmos. It comes out almost incredibly small. In other words, the amount of matter in space, compared with the volume of space it occupies, is almost infinitesimal. Lord Kelvin argues that ultimately it must be really infinitesimal (*Philosophical Magazine*, Aug., 1901, and Jan., 1902); that is to say that the volume of space is infinitely greater than the total bulk of matter which it contains. Otherwise the combined force of gravity—or at least the aggregate gravitational potential—on which the velocity generated in material bodies ultimately depends, would be far greater than observation shows it to be.

The whole visible universe, within a parallax of $\frac{1}{1000}$ second of arc, is estimated by Lord Kelvin as the equivalent of a thousand million of our suns; and this amount of matter, distributed as it is, would have an average density of 1.6×10^{-23} grammes per c.c. It is noteworthy how exceedingly small is this average or aggregate density of matter in the visible region of space. The estimated density of 10^{-23} c.g.s. means that the visible cosmos is as much rarer

THE ETHER OF SPACE

than a "vacuum" of a hundred millionths of an atmosphere, as that vacuum is itself rarer than lead.

It is because we have reason to assert that any ordinary mass of matter consists, like the cosmos, of separated particles, with great intervening distances in proportion to their size, that we are able to maintain that the aggregate density of ordinary stuff, such as water or lead, is very small compared with the continuous medium in which they exist, and of which all particles are supposed to be really composed. So that lead is to the ether, as regards density, very much as the "vacuum" above spoken of is to lead. The fundamental medium itself must be of uniform density everywhere, whether materialised or free.

VII

FURTHER EXPLANATIONS CONCERNING THE DENSITY AND ENERGY OF THE ETHER

A READER may suppose that in speaking of the immense density or massiveness of ether, and the absurdly small density or specific gravity of gross matter by comparison, I intend to signify that matter is a *rarefaction* of the ether. That, however, is not my intention. The view I advocate is that the ether is a perfect *continuum*, an absolute *plenum*, and that, therefore, no rarefaction is possible. The ether inside matter is just as dense as the ether outside, and no denser. A material unit—say, an electron—is only a peculiarity or singularity of some kind in the ether itself, which is of perfectly uniform density everywhere. What we "sense" as matter is an aggregate or grouping of an enormous number of such units.

How, then, can we say that matter is millions of times rarer or less substantial than the ether of which it is essentially composed? Those who feel any difficulty here, should bethink them-

THE ETHER OF SPACE

selves of what they mean by the average or aggregate density of any discontinuous system, such as a powder, or a gas, or a precipitate, or a snow-storm, or a cloud, or a milky way.

If it be urged that it is unfair to compare an obviously discrete assemblage like the stars, with an apparently continuous substance like air or lead, the answer is that it is entirely and accurately fair; since air, and every other known form of matter, is essentially an aggregate of particles, and since it is always their average density that we mean. We do not even know for certain their individual atomic density.

The phrase "specific gravity or density of a powder" is ambiguous. It may mean the specific gravity of the dry powder as it lies, like snow; or it may mean the specific gravity of the particles of which it is composed, like ice.

So also with regard to the density of matter; we might mean the density of the fundamental material of which its units are made—which would be ether; or we might, and in practice do, mean the density of the aggregate lump which we can see and handle; that is to say, of water or iron or lead, as the case may be.

In saying that the density of matter is small—I mean, of course, in the last, the usual, sense. In saying that the density of ether is great—I mean that the actual stuff of which these highly porous aggregates are composed is of immense, of

DENSITY AND ENERGY

well-nigh incredible, density. It is only another way of saying that the ultimate units of matter are few and far between—*i.e.*, that they are excessively small as compared with the distances between them; just as the planets of the solar system, or worlds in the sky, are few and far between—the intervening distances being enormous as compared with the portions of space actually occupied by lumps of matter.

It may be noted that it is not unreasonable to argue that the density of a *continuum* is necessarily greater than the density of any disconnected aggregate: certainly of any assemblage whose particles are actually composed of the material of the *continuum*. Because the former is "all there," everywhere, without break or intermittence of any kind; while the latter has gaps in it—it is here and there, but not everywhere.

Indeed, this very argument was used long ago by that notable genius Robert Hooke, and I quote a passage which Professor Poynting has discovered in his collected posthumous works and kindly copied out for me:—

"As for *matter*, that I conceive in its essence to be immutable, and its essence being expatiation determinate, it cannot be altered in its quantity, either by condensation or rarefaction; that is, there cannot be more or less of that power or reality, whatever it be, within the same expatiation or content; but every equal expatiation

THE ETHER OF SPACE

contains, is filled, or is an equal quantity of *materia;* and the densest or heaviest, or most powerful body in the world contains no more materia than that which we conceive to be the rarest, thinnest, lightest, or least powerful body of all; as gold for instance, and *æther*, or the substance that fills the cavity of an exhausted vessel, or cavity of the glass of a barometer above the quicksilver. Nay, as I shall afterwards prove, this cavity is more full, or a more dense body of æther, in the common sense or acceptation of the word, than gold is of gold, bulk for bulk; and that because the one—*viz.*, the mass of æther, is all æther: but the mass of gold, which we conceive, is not all gold; but there is an intermixture, and that vastly more than is commonly supposed, of æther with it; so that vacuity, as it is commonly thought, or erroneously supposed, is a more dense body than the gold as gold. But if we consider the whole content of the one with that of the other, within the same or equal quantity of expatiation, then are they both equally containing the *materia* or body."— *From the Posthumous Works of Robert Hooke, M.D., F.R.S.,* 1705, *pp.* 171–2 (*as copied in Memoir of Dalton, by Angus Smith*).

Newton's contemporaries did not excel in power of clear expression, as he himself did; but Professor Poynting interprets this singular attempt at utterance thus: "All space is filled

DENSITY AND ENERGY

with equally dense *materia*. Gold fills only a small fraction of the space assigned to it, and yet has a big mass. How much greater must be the total mass filling that space."

The tacit assumption here made is that the particles of the aggregate are all composed of one and the same continuous substance—practically that matter is made of ether; and that assumption, in Hooke's day, must have been only a speculation. But it is the kind of speculation which time is justifying, it is the kind of truth which we all feel to be in process of establishment now.[1]

We do not depend on that sort of argument, however; what we depend on is experimental measure of the mass, and mathematical estimate of the volume, of the electron. For calculation shows that however the mass be accounted for—whether electrostatically, or magnetically, or hydrodynamically—the estimate of ratio of mass to effective volume can differ only in a numerical coefficient, and cannot differ as regards order of magnitude. The only way out of this conclusion would be the discovery that the negative electron is not the real or the main matter-

---

[1] It does not seem to have been noticed that in Query 22, quoted in the Introduction to the present book, Newton seems to throw out a curious hint in this same direction, though he immediately abandons it again. He does not appear to have carefully *edited* his queries; probably they were published posthumously.

unit, but is only a subsidiary ingredient; whereas the main mass is the more bulky positive charge. That last hypothesis, however, is at present too vague to be useful. Moreover, the mass of such a charge would in that case be unexplained, and would need a further step; which would probably land us in much the same sort of etherial density as is involved in the estimate which I have based on the more familiar and tractable negative electron. (*See* Appendix 2.)

It may be said, Why assume any definite density for the ether at all? Why not assume that, as it is infinitely continuous, so it is infinitely dense —whatever that may mean—and that all its properties are infinite? This might be possible were it not for the velocity of light. By transmitting waves at a finite and measurable speed, the ether has given itself away, and has let in all the possibilities of calculation and numerical statement. Its properties are thereby exhibited as essentially finite—however infinite the whole extent of it may turn out to be. Parenthetically, we may remark that "gravitation" has not yet exhibited any similar kind of finite property; and that is why we know so little about it.

Etherial Energy.

Instead, then, of saying that the density of the ether is great, the clearest mode of expression

DENSITY AND ENERGY

is to say that the density of matter is small. Just as we can say that the density of the visible cosmos is small, although in individual lumps its density is comparable to that of iron or rock.

At the risk of repetition, I have explained this over again, because it is a matter on which confusion may easily arise. The really important thing about ether is not so much its density, considered in itself, as the energy which that density necessarily involves, on any kinetic theory of its elasticity. For it is not impossible —however hopeless it may seem now—that a modicum of that energy may some day be partially utilised.

Lord Kelvin's incipient kinetic theory of elasticity is a complicated matter, and I will only briefly enter upon it. But before doing so, I want to remove an objection which is sometimes felt, as to the fluid and easily permeable character of a medium of this great density— that is to say, as to the absence of friction or viscosity—the absence of resistance to bodies moving through it. As a matter of fact there is no necessary connection whatever between density and viscosity.

"Density" and "Viscosity" are entirely different things; and, if there is no fluid friction, a fluid may have any density you please without interposing any obstacle to constant velocity.

THE ETHER OF SPACE

To *acceleration* it does indeed oppose an obstacle, but that appears as essentially a part of the inertia or massiveness of the moving body. It contributes to its momentum; and, if the fluid is everywhere present, it is impossible to discriminate between, or to treat separately, that part of the inertia which belongs to the fluid displaced, and that part which belongs to the body moving through it—except by theory.

As for the elasticity of the ether, that is ascertainable at once from the speed at which it transmits waves. That speed—the velocity of light—is accurately known, 3×10^{10} centimeters per second. And the ratio of the elasticity or rigidity to the density is equal to the square of this speed; that is to say, the elasticity must be 9×10^{20} times the density; or, in other words, 10^{33} c.g.s. units. That is an immediate consequence of the estimate of density and the fact of the velocity of light; and if the density is admitted, the other cannot be contested.

But we must go on to ask, To what is this rigidity due? If the ether does not consist of parts, and if it is fluid, how can it possess the rigidity appropriate to a solid, so as to transmit transverse waves? To answer this we must fall back upon Lord Kelvin's kinetic theory of elasticity: that it must be due to rotational motion—intimate fine-grained motion throughout the whole etherial region—motion not of

DENSITY AND ENERGY

the nature of locomotion, but circulation in closed curves, returning upon itself—vortex motion of a kind far more finely grained than any waves of light or any atomic or even electronic structure.

Now if the elasticity of any medium is to be thus explained kinetically, it follows, as a necessary consequence, that the speed of this internal motion must be comparable to the speed of wave propagation; that is to say that the internal squirming circulation, to which every part of the ether is subject, must be carried on with a velocity of the same order of magnitude as the velocity of light.

This is the theory then—this theory of elasticity as dependent on motion—which, in combination with the estimate of density, makes the internal energy of the ether so gigantic. For in every cubic millimeter of space we have, according to this view, a mass equivalent to what, if it were matter, we should call a thousand tons, circulating internally, every part of it, with a velocity comparable to the velocity of light, and therefore containing—stored away in that small region of space—an amount of energy of the order 10^{29} ergs, or, what is the same thing, 3×10^{11} kilowatt centuries; which is otherwise expressible as equal to the energy of a million horsepower station working continuously for forty million years.

THE ETHER OF SPACE

SUMMARISED BRIEF STATEMENTS CONCERNING THE ETHER

(As communicated by the author to the British Association at Leicester, 1907.)

1. The theory that an electric charge must possess the equivalent of inertia was clearly established by J. J. Thomson in the *Philosophical Magazine* for April, 1881.

2. The discovery of masses smaller than atoms was made experimentally by J. J. Thomson, and communicated to Section A at Dover in 1899.

3. The thesis that the corpuscles so discovered consisted wholly of electric charges was sustained by many people, and was clinched by the experiments of Kaufmann in 1902.

4. The concentration of the ionic charge, required to give the observed corpuscular inertia, can be easily calculated; and consequently the size of the electric nucleus, or electron, is known.

5. The old perception that a magnetic field is kinetic has been developed by Kelvin, Heaviside, FitzGerald, Hicks, and Larmor, most of whom have treated it as a flow along magnetic lines; though it may also, perhaps equally well, be regarded as a flow perpendicular to them and along the Poynting vector. The former doctrine is sustained by Larmor, as in accordance with the

DENSITY AND ENERGY

principle of Least Action, and with the absolutely stationary character of the ether as a whole; the latter view appears to be more consistent with the theories of J. J. Thomson.

6. A charge in motion is well known to be surrounded by a magnetic field; and the energy of the motion can be expressed in terms of the energy of this concomitant field—which again must be accounted as the kinetic energy of ethereous flow.

7. Putting these things together, and considering the ether as essentially incompressible—on the strength of the Cavendish electric experiment, the facts of gravitation, and the general idea of a connecting continuous medium—the author reckons that to deal with the ether dynamically it must be treated as having a density of the order 10^{12} grammes per cubic centimeter. (*See* Appendix 2.)

8. The existence of transverse waves in the interior of a fluid can only be explained on gyrostatic principles—*i.e.*, by the kinetic or rotational elasticity of Lord Kelvin. And the internal circulatory speed of the intrinsic motion of such a fluid must be comparable with the velocity with which such waves are transmitted.

9. Putting these things together, it follows that the intrinsic or constitutional vortex energy of the ether must be of the order 10^{33} ergs per cubic centimeter.

THE ETHER OF SPACE

Conclusion.—Thus every cubic millimeter of the universal ether of space must possess the equivalent of a thousand tons, and every part of it must be squirming internally with the velocity of light.

VIII

ETHER AND MATTER

THE MECHANICAL NECESSITY FOR A CONTINUOUS MEDIUM FILLING SPACE

IN this chapter I propose to summarise in simple and consecutive form most of the arguments already used. Thirty years ago Clerk-Maxwell gave to the Royal Institution of Great Britain a remarkable address on "Action at a Distance." It is reported in the Journal R.I., Vol. VII, and to it I would direct attention. Most natural philosophers hold, and have held, that action at a distance across empty space is impossible; in other words, that matter cannot act where it is not, but only where it is. The question "Where is it?" is a further question that may demand attention and require more than a superficial answer. For it can be argued on the hydro-dynamic or vortex theory of matter, as well as on the electrical theory, that every atom of matter has a universal though nearly infinitesimal prevalence, and extends everywhere; since there is no definite sharp

THE ETHER OF SPACE

boundary or limiting periphery to the region disturbed by its existence. The lines of force of an isolated electric charge extend throughout illimitable space. And though a charge of opposite sign will curve and concentrate them, yet it is possible to deal with both charges, by the method of superposition, as if they each existed separately without the other.

In that case, therefore, however far they reach, such nuclei clearly exert no "action at a distance" in the technical sense.

Some philosophers have reason to suppose that mind can act directly on mind without intervening mechanism—and sometimes that has been spoken of as genuine action at a distance; but no proper conception or physical model can be made of such a process, nor is it clear that "space" and "distance" have any particular meaning in the region of psychology. The links between mind and mind may be something quite other than physical proximity; and in denying action at a distance across empty space I am not denying telepathy or other activities of a nonphysical kind. For although brain disturbance is certainly physical, and is an essential concomitant of mental action whether of the sending or receiving variety, yet we know from the case of heat that a material movement can be excited in one place at the expense of corresponding movement in another, without any similar

ETHER AND MATTER

kind of transmission or material connection between the two places: the thing that travels across vacuum is not heat.

In all cases where physical motion is involved, however, I would have a medium sought for. It may not be matter, but it must be something; there must be a connecting link of some kind, or the transference cannot occur. There can be no attraction across really empty space. And even when a material link exists, so that the connection is obvious, the explanation is not complete; for when the mechanism of attraction is understood, it will be found that a body really only moves because it is pushed by something from behind. The essential force in nature is the *vis a tergo*. So when we have found the "traces," or discovered the connecting thread, we still run up against the word "cohesion"; and we ought to be exercised in our minds as to its ultimate meaning. Why the whole of a rod should follow, when one end is pulled, is a matter requiring explanation; and the only explanation that can be given involves, in some form or other, a continuous medium connecting the discrete and separated particles or atoms of matter.

When a steel spring is bent or distorted, what is it that is really strained? Not the atoms—the atoms are only displaced; it is the connecting links that are strained—the connecting medium—the ether. Distortion of a spring is

THE ETHER OF SPACE

really distortion of the ether. All stress exists in the ether. Matter can only be moved. Contact does not exist between the atoms of matter as we know them; it is doubtful if a piece of matter ever touches another piece, any more than a comet touches the sun when it appears to rebound from it; but the atoms are connected, as the comet and the sun are connected, by a continuous *plenum* without break or discontinuity of any kind. Matter acts on matter only through the ether. But whether matter is a thing utterly distinct and separate from the ether, or whether it is a specifically modified portion of it—modified in such a way as to be susceptible of locomotion and yet continuous with all the rest of the ether, which can be said to extend everywhere far beyond the bounds of the modified and tangible portion—are questions demanding, and I may say in process of receiving, answers.

Every such answer involves some view of the universal and possibly infinite uniform omnipresent connecting medium, the Ether of Space.

It has been said, somewhat sarcastically, that the ether was made in England. The statement is only an exaggeration of the truth. I might even urge that it has been largely constructed in the Royal Institution; for, I will summarise now the chief lines of evidence on which its

ETHER AND MATTER

existence is believed in, and our knowledge of it is based.

First of all, Newton recognised the need of a medium for explaining gravitation. In his "Optical Queries" he shows that if the pressure of this medium is less in the neighbourhood of dense bodies than at great distances from them, dense bodies will be driven toward each other; and that if the diminution of pressure is inversely as the distance from the dense body, the law of force will be the inverse square law of gravitation.

All that is required, therefore, to explain gravity, is a diminution of pressure, or increase of tension, caused by the formation of a matter unit—that is to say of an electron or corpuscle. And although we do not yet know what an electron is—whether it be a strain centre, or what kind of singularity in the ether it may be—there is no difficulty in supposing that a slight, almost infinitesimal, strain or attempted rarefaction should be produced in the ether whenever an electron comes into being—to be relaxed again only on its resolution and destruction. Strictly speaking it is not a real *strain*, but only a "stress"; since there can be no actual *yield*, but only a pull or tension, extending in all directions toward infinity.

The tension required per unit of matter is almost ludicrously small, and yet in the aggre-

THE ETHER OF SPACE

gate, near such a body as a planet, it becomes enormous.

The force with which the moon is held in its orbit would be great enough to tear asunder a steel rod four hundred miles thick, with a tenacity of 30 tons per square inch; so that if the moon and earth were connected by steel instead of by gravity, a forest of pillars would be necessary to whirl the system once a month round their common centre of gravity. Such a force necessarily implies enormous tension or pressure in the medium. Maxwell calculates that the gravitational stress near the earth, which we must suppose to exist in the invisible medium, is 3000 times greater than what the strongest steel could stand; and near the sun it should be 2500 times as great as that.

The question has arisen in my mind, whether, if the whole sensible universe—estimated by Lord Kelvin as equivalent to about a thousand million suns—were all concentrated in one body of specifiable density,[1] the stress would not be so great as to produce a tendency toward etherial disruption; which would result in a disintegrating explosion, and a scattering of the particles once more as an enormous nebula and other fragments into the depths of space. For the

[1] On doing the arithmetic, however, I find the necessary concentration absurdly great, showing that such a mass is quite insufficient. (*See* Appendix 1.)

ETHER AND MATTER

tension would be a maximum in the interior of such a mass; and, if it rose to the value 10^{33} dynes per square centimeter, something would have to happen. I do not suppose that this can be the reason, but one would think there must be *some* reason, for the scattered condition of gravitative matter.

Too little is known, however, about the mechanism of gravitation to enable us to adduce it as the strongest argument in support of the existence of an ether. The oldest valid and conclusive requisition of an ethereous medium depends on the wave theory of light, one of the founders of which was the Royal Institution Professor of Natural Philosophy at the beginning of last century, Dr. Thomas Young.

No ordinary matter is capable of transmitting the undulations or tremors that we call light. The speed at which they go, the kind of undulation, and the facility with which they go through vacuum, forbid this.

So clearly and universally has it been perceived that waves must be waves of something —something distinct from ordinary matter— that Lord Salisbury, in his presidential address to the British Association at Oxford, criticised the ether as little more than a nominative case to the verb to undulate. It is truly *that*, though it is also truly more than that; but to illustrate that luminiferous aspect of it, I will quote a

THE ETHER OF SPACE

paragraph from the lecture of Clerk-Maxwell's to which I have already referred:—

"The vast interplanetary and interstellar regions will no longer be regarded as waste places in the universe, which the Creator has not seen fit to fill with the symbols of the manifold order of His kingdom. We shall find them to be already full of this wonderful medium; so full, that no human power can remove it from the smallest portion of space, or produce the slightest flaw in its infinite continuity. It extends unbroken from star to star; and when a molecule of hydrogen vibrates in the dog-star, the medium receives the impulses of these vibrations, and after carrying them in its immense bosom for several years, delivers them, in due course, regular order, and full tale, into the spectroscope of Mr. Huggins, at Tulse Hill."

This will suffice to emphasise the fact that the eye is truly an etherial sense-organ—the only one which we possess, the only mode by which the ether is enabled to appeal to us; and that the detection of tremors in this medium—the perception of the direction in which they go, and some inference as to the quality of the object which has emitted them—cover all that we mean by "sight" and "seeing."

I pass, then, to another function: the electric and magnetic phenomena displayed by the ether; and on this I will only permit myself a very short

ETHER AND MATTER

quotation from the writings of Faraday, whose whole life may be said to have been directed toward a better understanding of these ethereous phenomena. Indeed, the statue in the entrance hall of the Royal Institution, Albemarle Street, may be considered as the statue of the discoverer of the electric and magnetic properties of the Ether of Space.

Faraday conjectured that the same medium which is concerned in the propagation of light might also be the agent in electromagnetic phenomena. "For my own part," he says, "considering the relation of a vacuum to the magnetic force, and the general character of magnetic phenomena external to the magnet, I am much more inclined to the notion that in the transmission of the force there is such an action, external to the magnet, than that the effects are merely attraction and repulsion at a distance. Such an action may be a function of the æther; for it is not unlikely that, if there be an æther, it should have other uses than simply the conveyance of radiation."

This conjecture has been amply strengthened by subsequent investigations.

One more function is now being discovered; the ether is being found to constitute matter—an immensely interesting topic, on which there are many active workers at the present time. I will make a brief quotation from Prof. Sir J. J.

Thomson, where he summarises the conclusion which we all see looming before us, though it has not yet been completely attained, and would not by all be similarly expressed:—

"The *whole* mass of any body is just the mass of ether surrounding the body which is carried along by the Faraday tubes associated with the atoms of the body. In fact, all mass is mass of the ether; all momentum, momentum of the ether; and all kinetic energy, kinetic energy of the ether. This view, it should be said, requires the density of the ether to be immensely greater than that of any known substance."

Yes, far denser—so dense that matter by comparison is like gossamer, or a filmy imperceptible mist, or a milky way. Not unreal or unimportant—a cobweb is not unreal, nor to certain creatures is it unimportant, but it cannot be said to be massive or dense; and matter, even platinum, is not dense when compared with the ether. Not till last year, however, did I realise what the density of the ether must really be,[1] compared with that modification of it which appeals to our senses as matter, and which for that reason engrosses our attention.

Is there any other function possessed by the ether, which, though not yet discovered, may lie within the bounds of possibility for future dis-

---

[1] See Lodge, *Philosophical Magazine*, April, 1907. Also Appendix 2, below.

ETHER AND MATTER

covery? I believe there is, but it is too speculative to refer to, beyond saying that it has been urged as probable by the authors of *The Unseen Universe*, and has been thus tentatively referred to by Clerk-Maxwell:—

"Whether this vast homogeneous expanse of isotropic matter is fitted not only to be a medium of physical interaction between distant bodies, and to fulfil other physical functions of which, perhaps, we have as yet no conception, but also ... to constitute the material organism of beings exercising functions of life and mind as high or higher than ours are at present—is a question far transcending the limits of physical speculation."

And there for the present I leave that aspect of the subject.

Ether and Matter.

I shall now attempt to illustrate some relations between ether and matter.

The question is often asked, Is ether material? This is largely a question of words and convenience. Undoubtedly, the ether belongs to the material or physical universe, but it is not ordinary matter. I should prefer to say it is not "matter" at all. It may be the substance or substratum or material of which matter is composed, but it would be confusing and inconvenient not to be able to discriminate between matter on the one hand and ether on the

THE ETHER OF SPACE

other. If you tie a knot on a bit of string, the knot is composed of string, but the string is not composed of knots. If you have a smoke or vortex ring in the air, the vortex-ring is made of air, but the atmosphere is not a vortex-ring; and it would be only confusing to say that it was.

The essential distinction between matter and ether is that matter *moves*, in the sense that it has the property of locomotion and can effect impact and bombardment; while ether is *strained*, and has the property of exerting stress and recoil. All potential energy exists in the ether. It may vibrate, and it may rotate, but as regards locomotion it is stationary—the most stationary body we know: absolutely stationary, so to speak; our standard of rest.

All that we ourselves can effect, in the material universe, is to alter the motion and configuration of masses of matter; we can move matter by our muscles, and that is all we can do directly: everything else is indirect.

But now comes the question, How is it possible for matter to be composed of ether? How is it possible for a solid to be made out of fluid? A solid possesses the properties of rigidity, impenetrability, elasticity, and such like; how can these be imitated by a perfect fluid such as the ether must be?

The answer is, They can be imitated by *a fluid in motion;* a statement which we make with

ETHER AND MATTER

confidence as the result of a great part of Lord Kelvin's work.

It may be illustrated by a few experiments.

A wheel of spokes, transparent or permeable when stationary, becomes opaque when revolving, so that a ball thrown against it does not go through, but rebounds. The motion only affects permeability to matter; transparency to light is unaffected.

A silk cord hanging from a pulley becomes rigid and viscous when put into rapid motion; and pulses or waves which may be generated on the cord travel along it with a speed equal to its own velocity, whatever that velocity may be, so that they appear to stand still. This is a genuine case of kinetic rigidity; and the fact that the wave-transmission velocity is equal to the rotatory speed of the material, is typical and important, for in all cases of kinetic elasticity these two velocities are of the same order of magnitude.

A flexible chain, set spinning, can stand up on end while the motion continues.

A jet of water at sufficient speed can be struck with a hammer, and resists being cut with a sword.

A spinning disk of paper becomes elastic like flexible metal, and can act like a circular saw. Sir William White tells me that in naval construction steel plates are cut by a rapidly revolving disk of soft iron.

THE ETHER OF SPACE

A vortex-ring, ejected from an elliptical orifice, oscillates about the stable circular form, as an india-rubber ring would do; thus furnishing a beautiful example of kinetic elasticity, and showing us clearly a fluid displaying some of the properties of a solid.

A still further example is Lord Kelvin's model of a spring balance, made of nothing but rigid bodies in spinning motion.[1] This arrangement utilises the processional movement of balanced gyrostats—concealed in a case and supporting a book—to imitate the behaviour of a spiral spring, if it were used to support the same book.

If the ether can be set spinning, therefore, we may have some hope of making it imitate the properties of matter, or even of constructing matter by its aid. But *how* are we to spin the ether? Matter alone seems to have no grip of it. As already described, I have spun steel disks, a yard in diameter, 4000 times a minute, have sent light round and round between them, and tested carefully for the slightest effect on the ether. Not the slightest effect was perceptible. We cannot spin ether mechanically.

But we can vibrate it electrically; and every source of radiation does that. An electrical charge, in sufficiently rapid vibration, is the only source of ether-waves that we know; and

[1] Address to Section A of British Association at Montreal, 1884.

ETHER AND MATTER

if an electric charge is suddenly stopped, it generates the pulses known as X-rays, as the result of the collision. Not speed, but sudden change of speed, is the necessary condition for generating waves in the ether by electricity.

We can also infer some kind of rotary motion in the ether; though we have no such obvious means of detecting the spin as is furnished by vision for detecting some kinds of vibration. Rotation is supposed to exist whenever we put a charge into the neighbourhood of a magnetic pole. Round the line joining the two, the ether is spinning like a top. I do not say it is spinning fast: that is a question of its density; it is, in fact, spinning with excessive slowness, but it is spinning with a definite moment of momentum. J. J. Thomson's theory makes its moment of momentum exactly equal to $e\, m$, the product of *charge* and *pole;* the charge being measured electrostatically and the pole magnetically.

How can this be shown experimentally? Suppose we had a spinning top enclosed in a case, so that the spin was unrecognisable by ordinary means—it could be detected by its gyrostatic behaviour to force. If allowed to "precess" it will respond by moving perpendicularly to a deflecting force. So it is with the charge and the magnetic pole. Try to move the charge suddenly, and it immediately sets off at right angles. A moving charge is a current, and the

THE ETHER OF SPACE

pole and the current try to revolve round one another—a fact which may be regarded as exhibiting a true gyrostatic action due to the otherwise unrecognisable etherial spin. The fact of such magnetic rotation was discovered by Faraday.

I know that it is usually worked out in another way, in terms of lines of force and the rest of the circuit; but I am thinking of a current as a stream of projected charges; and no one way of regarding such a matter is likely to exhaust the truth, or to exclude other modes which are equally valid. Anyhow, in whatever way it is regarded, it is an example of the three rectangular vectors.

The three vectors at right angles to each other, which may be labelled Current, Magnetism, and Motion respectively, or more generally E, H, and V, represent the quite fundamental relation between ether and matter, and constitute the link between Electricity, Magnetism, and Mechanics. Where any two of these are present, the third is a necessary consequence. This principle is the basis of all dynamos, of electric motors, of light, of telegraphy, and of most other things. Indeed, it is a question whether it does not underlie everything that we know in the whole of the physical sciences; and whether it is not the basis of our conception of the three dimensions of space.

Lastly, we have the fundamental property of

ETHER AND MATTER

matter called *inertia*, which can, to a certain extent, be explained electromagnetically, provided the ethereous density is granted as of the order 10^{12} grammes per cubic centimeter. The elasticity of the ether would then have to be of the order 10^{33} c.g.s.; and if this is due to intrinsic turbulence, the speed of the whirling or rotational elasticity must be of the same order as the velocity of light. This follows hydrodynamically; in the same sort of way as the speed at which a pulse travels on a flexible running endless cord, whose tension is entirely due to the centrifugal force of the motion, is precisely equal to the velocity of the cord itself. And so, on our present view, the intrinsic energy of constitution of the ether is incredibly and portentously great; every cubic millimeter of space possessing what, if it were matter, would be a mass of a thousand tons, and an energy equivalent to the output of a million-horse-power-station for 40 million years.

The universe we are living in is an extraordinary one; and our investigation of it has only just begun. We know that matter has a psychical significance, since it can constitute *brain*, which links together the physical and the psychical worlds. If any one thinks that the ether, with all its massiveness and energy, has probably no psychical significance, I find myself unable to agree with him.

IX

STRENGTH OF THE ETHER

TO show that the ether cannot be the slight and rarefied substance which at one time, and indeed until quite lately, it was thought to be, it is useful to remember that not only has it to be the vehicle of light and the medium of all electric and magnetic influence, but also that it has to transmit the tremendous forces of gravitation.

Among small bodies gravitational forces are slight, and are altogether exceeded by magnetic and electric or chemical forces. Indeed, gravitational attraction between bodies of a certain smallness can be more than counterbalanced even by the pressure which their mutual radiation exerts—almost infinitesimal though that is; so that, as a matter of fact, small enough bodies of any warmth will repel each other unless they are in an enclosure of constant temperature—*i.e.*, unless the radiation pressure upon them is uniform all round.

The size at which radiation repulsion overbalances gravitational attraction, for equal

STRENGTH OF THE ETHER

spheres, depends on the temperature of the spheres and on their density; but at the ordinary temperature to which we are accustomed, say 60° Fahrenheit or thereabouts, equality between the two forces will obtain for two wooden spheres in space if each is about a foot in diameter; according to Professor Poynting's data (*Philosophical Transactions*, Vol. 202, p. 541). For smaller or hotter bodies, radiation-repulsion overpowers mutual gravitation; and it increases with the fourth power of their absolute temperature. The gravitational attractive force between particles is exceedingly small; and that between two atoms or two electrons is negligibly small, even though they be within molecular distance of each other.

For instance, two atoms of, say, gold, at molecular distance, attract each other gravitationally with a force of the order

$$\gamma \frac{10^{-22} \times 10^{-22}}{(10^{-8})^2} = \frac{10^{-44}}{10^{-16}} \times 10^{-7} = 10^{35} \text{ dyne};$$

which would cause no perceptible acceleration at all.

The gravitational attraction of two electrons at the same distance is the forty-thousand-millionth part of this, and so one would think must be entirely negligible. And yet it is to the aggregate attraction of myriads of such bodies that the resultant force of attraction is due—

THE ETHER OF SPACE

a force which is felt over millions of miles. The force is not only felt indeed, but must be reckoned as one of prodigious magnitude.

When dealing with bodies of astronomical size, the force of gravitation overpowers all other forces; and all electric and magnetic attractions sink by comparison into insignificance.

These immense forces must be transmitted by the ether, and it is instructive to consider their amount.

SOME ASTRONOMICAL FORCES WHICH THE ETHER HAS TO TRANSMIT.

Arithmetical Calculation of the Pull of the Earth on the Moon.

The mass of the earth is 6000 trillion (6×10^{21}) tons. The mass of the moon is $\frac{1}{80}$th that of the earth. Terrestrial gravity at the moon's distance (which is 60 earth radii) must be reduced in the ratio $1 : 60^2$; that is, it must be $\frac{1}{3600}$th of what it is here.

Consequently the pull of the earth on the moon is

$$\frac{6 \times 10^{21}}{80 \times 3600} \text{ tons weight.}$$

A pillar of steel which could transmit this force, provided it could sustain a tension of 40 tons to the square inch, would have a diameter of about 400 miles; as stated in the text, page 112.

STRENGTH OF THE ETHER

If this force were to be transmitted by a forest of weightless pillars each a square foot in cross-section, with a tension of 30 tons to the square inch throughout, there would have to be 5 million million of them.

Arithmetical Calculation of the Pull of the Sun on the Earth.

The mass of the earth is 6×10^{21} tons. The intensity of solar gravity at the sun's surface is 25 times ordinary terrestrial gravity.

At the earth's distance, which is nearly 200 solar radii, solar gravity will be reduced in the ratio of $1:200$ squared.

Hence the force exerted by the sun on the earth is

$$\frac{25 \times 6 \times 10^{21}}{(200)^2} \text{ tons weight.}$$

That is to say, it is approximately equal to the weight of 37×10^{17} ordinary tons upon the earth's surface.

Now steel may readily be found which can stand a load of 37 tons to every square inch of cross-section. The cross-section of a bar of such steel, competent to transmit the sun's pull to the earth, would therefore have to be

10^{17} square inches;
or, say, 700×10^{12} square feet.

THE ETHER OF SPACE

And this is equivalent to a million million round rods or pillars each 30 feet in diameter.

Hence the statement in the text (page 26) is well within the mark.

The Pull of the Earth on the Sun.

The pull of the earth on the sun is, of course, equal and opposite to the pull of the sun on the earth, which has just been calculated; but it furnishes another mode of arriving at the result, and may be regarded as involving simpler data— *i.e.*, data more generally known. All we need say is the following:—

The mass of the Sun is 316,000 times that of the Earth.

The mean distance of the sun is, say, 23,000 earth radii.

Hence the weight or pull of the sun by the earth is

$$\frac{316000}{(23000)^2} \times 6 \times 10^{21} \text{ tons weight.}$$

In other words, it is approximately equal to the ordinary commercial weight of 36×10^{17} tons, as already calculated.

The Centripetal Force acting on the Earth.

Yet another method of calculating the sun's pull is to express it in terms of the centrifugal force of the earth; namely, its mass, multiplied

STRENGTH OF THE ETHER

by the square of its angular velocity, multiplied by the radius of its orbit; that is to say,

$$F = M \left(\frac{2\pi}{T}\right)^2 r$$

where T is the length of a year.

The process of evaluating this is instructive, owing to the manipulation of units which it involves:—

$$F = 6 \times 10^{21} \text{ tons} \times \frac{4\pi^2 \times 92 \times 10^6 \text{ miles}}{(365\tfrac{1}{4} \text{ days})^2}$$

which of course is a mass multiplied by an acceleration. The acceleration is—

$$\frac{40 \times 92 \times 10^6}{133300 \times (24)^2} \text{ miles per hour per hour}$$

$$= \frac{3680 \times 10^6 \times 5280}{133300 \times 576 \times (3600)^2} \text{ feet per sec. per sec.}$$

$$= \frac{115 \times 5280}{133300 \times 576 \times 12.96} \times 32 \text{ feet per sec. per sec.}$$

$$= \frac{g}{1640}$$

Hence the Force of attraction is that which, applied to the earth's mass, produces in it an acceleration equal to the $\frac{1}{1640}$th part of what ordinary terrestrial gravity can produce in falling bodies; or

$$F = 6 \times 10^{21} \text{ tons} \times \frac{g}{1640}$$

$$= \frac{6}{1640} \times 10^{21} \text{ tons weight;}$$

THE ETHER OF SPACE

which is the ordinary weight of 37×10^{17} tons, as before.

The slight numerical discrepancy between the above results is of course due to the approximate character of the data selected, which are taken in round numbers as quite sufficient for purposes of illustration.

If we imagine the force applied to the earth by a forest of round rods, one for every square foot of the earth's surface—*i.e.*, of the projected earth's hemisphere or area of equatorial plane—the force transmitted by each would have to be 2700 tons; and therefore, if of 30-ton steel, they would each have to be eleven inches in diameter, or nearly in contact, all over the earth.

Pull of a Planet on the Earth.

While we are on the subject, it seems interesting to record the fact that the pull of any planet on the earth, even Neptune, distant though it is, is still a gigantic force. The pull of Neptune is $\frac{1}{30000}$th of the sun's pull; *i.e.*, it is 18 billion tons weight.

Pull of a Star on the Earth.

On the other hand, the pull of a fixed star, like Sirius—say a star, for example, which is 20 times the mass of the sun and 24 light years distant—is comparatively very small.

STRENGTH OF THE ETHER

It is easily found by dividing 20 times the sun's pull by the squared ratio of 24 years to 8 minutes; and it comes out as 30 million tons weight.

Such a force is able to produce no perceptible effect. The acceleration it causes in the earth and the whole solar system, at its present speed through space, is only able to curve the path with a radius of curvature of length thirty thousand times the distance of the star.

Force required to hold together the Components of some Double Stars.

But it is not to be supposed that the transmission of any of these forces gives the ether the slightest trouble, or strains it to anywhere near the limits of its capacity. Such forces must be transmitted with perfect ease, for there are plenty of cases where the force of gravitation is vastly greater than that. In the case of double stars, for instance, two suns are whirling round each other; and some of them are whirling remarkably fast. In such cases the force holding the components together must be enormous.

Perhaps the most striking case, for which we have substantially accurate data, is the star β Aurigæ; which, during the general spectroscopic survey of the heavens undertaken by Professor Pickering, of Harvard, in connection with the Draper Memorial, was discovered to show a spectrum with the lines some days double and alternate days single. Clearly it must consist of a pair of luminous objects revolving in a plane approximately containing the line of vision; the revolution being completed every four days. For the lines will then be optically displaced by the motion during part of the orbit—those of the advancing body to the right, those of the receding body to the left—while in that part of the orbit which lies athwart the direction of vision, the spectrum lines will return to their proper places, opening out again to a maximum, in the opposite direction, at the next quadrant.

The amount of displacement can be roughly estimated, enabling us to calculate the speed with which the sources of light were moving.

Professor Pickering, in a brief statement in *Nature*, Vol. XLI, page 403, 1889, says that the velocity amounts to about 150 miles per second, and that it is roughly the same for both components.

Taking these data:—
>Equality and uniformity of speeds, 150 miles per second each,
>Period 4 days—

we have all the data necessary to determine the masses; and likewise the gravitative pull between them. For the star must consist of two

STRENGTH OF THE ETHER

equal bodies, revolving about a common centre of gravity midway between them, in nearly circular orbits.

The speed and period together easily give the radius of the circular orbit as about 8 million miles.

Equating centrifugal and centripetal forces

$$\frac{m v^2}{r} = \gamma \frac{m^2}{(2r)^2}$$

and comparing the value of $4r^3/T^2$ so obtained with the r^3/T^2 of the earth, we find the mass of each body must be about 30,000 times that of the earth, or about $\frac{1}{10}$th that of the sun.

(This is treating them as spheres, though they must really be pulled into decidedly prolate shapes. Indeed it may seem surprising that the further portions can keep up with the nearer portions as they revolve. If they are of something like solar density their diameter will be comparable to half a million miles, and the natural periods of their near and far portions will differ in the ratio $(\frac{17}{16})^{3/2} = 1.1$ approximately. Tenacity could not hold the parts together, but gravitational coherence would.)

This, however, is a digression. Let us continue the calculation of the gravitative pull.

We have masses of $3 \times 10^4 \times 6 \times 10^{21}$ tons, re-

THE ETHER OF SPACE

volving with angular velocity $2\pi \div 4$ days, in a circle of radius 8×10^6 miles.

Consequently the centripetal acceleration is $\frac{4\pi^2 \times 8 \times 10^6}{16}$ miles per day per day; which comes out $\frac{32}{2.2}$ ft. per sec. per sec., or nearly half ordinary terrestrial gravity.

Consequently the pull between the two components of the double star β Aurigæ is

$$\frac{g}{2.2} \times 18 \times 10^{25} \text{ tons,}$$

or equal to the weight of

$$80 \times 10^{24} \text{ tons on the earth,}$$

which is more than twenty million times as great as is the pull between the earth and our sun.

Simple calculations such as these could have been made at any time; there is nothing novel about them, as there is about the estimate of the ether's density and vast intrinsic energy, in Chapters VI and VII. But then there is nothing hypothetical or uncertain about them either; they are certain and definite: whereas it may be thought there is something doubtful about the newer contentions which involve consideration of the mass and size of electrons and of the uniform and incompressible character of ethereal constitution. Even the idea of "massiveness"

STRENGTH OF THE ETHER

as applied to the ether involves an element of uncertainty or of figurativeness; because until we know more about ether's peculiar nature (if it is peculiar), we have to deal with it in accordance with material analogies, and must specify its massiveness as that which would have to be possessed by it if it fulfilled its functions and yet were anything like ordinary matter. It cannot really *be* ordinary matter, because ordinary matter is definitely differentiated from it, and is presumably composed of it; but the inertia of ordinary matter, however it be electrically or magnetically explained, must in the last resort depend on something parentally akin to inertia in the fundamental substance which fills space. And this it is which we have attempted in Chapters VI and VII to evaluate and to express in the soberest terms possible.

X

GENERAL THEORY OF ABERRATION

IN Chapter III the subject of Aberration was treated in a simple and geometrical manner, but it is now time to deal with it more generally. And to do this compactly I must be content in the greater part of this chapter to appeal chiefly to physicists.

The following general statements concerning aberration can be made:—

1. A ray of light in clear space is straight, whatever the motion of the medium, unless eddies exist; in other words, no irrotational disturbance of ether can deflect a ray.

2. But if the observer is in motion, the apparent ray will not be the true ray, and his line of vision will not truly indicate the direction of an object.

3. In a stationary ether the ray coincides with wave-normal. In a moving ether the ray and wave-normal enclose an aberration angle ϵ, such that $\sin \epsilon = v/V$, the ratio of the ether speed to the light speed.

ABERRATION THEORY

4. In all cases the line of vision depends on motion of the observer, and on that alone. If the observer is stationary, his line of vision is a ray. If he moves at the same rate as the ether, his line of vision is a wave-normal.

5. Line of vision depends not at all on the motion of the ether, so long as it has a velocity-potential. Hence, if this condition is satisfied the theory of aberration is quite simple.

General Statement as to Negative Results in the Subject.

It is noteworthy that almost all the observations which have been made with negative results as to the effect of the earth's orbital motion on the ether are equally consistent with complete connection and complete independence between ether and matter. If there is complete connection, the ether near the earth is relatively stagnant, and negative terrestrial results are natural. If there is complete independence, the ether is either absolutely stationary or has a velocity-potential, and the negative results are, as has been shown, thereby explained. Direct experiment on the subject of etherial viscosity proves that that is either really or approximately zero, and substantiates the "independence" explanation.

THE ETHER OF SPACE

Definition of a Ray.

A ray signifies the path of a definite or identical portion of radiation energy—the direction of energy-flux. In other words, it may be considered as the path of a labelled disturbance; for it is some special feature which enables an eye to fix direction: it is that which determines the line of collimation of a telescope.

Now in order that a disturbance from A may reach B, it is necessary that adjacent elements of a wave front at A shall arrive at B in the same phase; hence, the path by which a disturbance travels must satisfy this condition from point to point. This condition will be satisfied if the time of journey down a ray and down all infinitesimally differing paths is the same.

The equation to a ray is therefore contained in the statement that the time taken by light to traverse it is a minimum; or

$$\int_A^B \frac{ds}{V} = \text{minimum}$$

If the medium, instead of being stationary, is drifting with the velocity v, at angle θ to the ray, we must substitute for V the modified velocity $V \cos \epsilon + v \cos \theta$; and so the function

ABERRATION THEORY

that has to be a minimum, in order to give the path of a ray in a moving medium, is

$$\text{Time of journey} = \int_A^B \frac{ds}{V(\cos \epsilon + a \cos \theta)}$$

$$= \int_A^B \frac{V \cos \epsilon - v \cos \theta}{V^2 (1 - a^2)} ds = \text{minimum}$$

where a is the ratio v/V.

Path of Ray, and Time of Journey, through an Irrotationally Moving Medium.

Writing a velocity-potential ϕ in the above equation to a ray, that is putting

$$v \cos \theta = \frac{\delta \phi}{\delta s},$$

and ignoring possible variations in the minute correction factor $1 - a^2$ between the points A and B, it becomes

$$\text{Time of journey} = \int_A^B \frac{\cos \epsilon}{1 - a^2} \cdot \frac{ds}{V} - \frac{\phi_B - \phi_A}{V^2 (1 - a^2)} = \text{minimum}.$$

Now the second term depends only on end points, and therefore has no effect on path. The

THE ETHER OF SPACE

first term contains only the second power of aberration magnitude; and hence it has much the same value as if everything were stationary. A ray that was straight will remain straight in spite of motion. Whatever shape it had, that it will retain.

Only $\cos \epsilon$, and variations in a^2, can produce any effect on path; and effects so produced must be very small, since the value of $\cos \epsilon$ is

$$\sqrt{(1 - a^2 \sin^2 \theta)}.$$

A second-order effect on direction may therefore be produced by irrotational motion, but not a first-order effect. A similar statement applies to the time of journey round any closed periphery.

Michelson's Experiment.

We conclude, therefore, that general etherial drift does not affect either the path of a ray, or the time of its journey to and fro, or round a complete contour, to any important extent. But that taking second-order quantities into account, the time of going to and fro in any direction inclined at angle θ to a constant drift is, from the above expression,

$$T_1 + T_2 = \frac{2 T \cos \epsilon}{1 - a^2} = \frac{\sqrt{(1 - a^2 \sin^2 \theta)}}{1 - a^2} \times 2T,$$

where $2T$ is the ordinary time of the double journey without any drift.

ABERRATION THEORY

Hence some slight modification of interference effects by reason of drift would seem to be possible; since the time of a to and fro light-journey depends subordinately on the inclination of ray to drift.

The above expression applies to Michelson's remarkable experiment[1] of sending a split beam to and fro, half along and half across the line of the earth's motion; and is, in fact, a theory of it. There ought to be an effect due to the difference between $\theta=0$ and $\theta=90°$. But none can be detected. Hence, either something else happens, or the ether near the earth is dragged with it so as not to stream through our instruments.

Alternative Explanation.

But if the ether is dragged along near moving matter, it behaves like a viscous fluid, and all idea of a velocity-potential must be abandoned. This would complicate the theory of aberration (pp. 47 and 64), and moreover is dead against the experimental evidence described in Chapter V.

The negative result of Mr. Michelson's is, however, explicable in another way—namely, by the FitzGerald-Lorentz theory that the linear dimensions of bodies are a function of their motion through the ether. And such an effect it is reasonable to expect; since, if cohesion forces are

[1] *Philosophical Magazine*, December, 1887.

THE ETHER OF SPACE

electrical, they must be affected by motion, to a known and calculable amount, depending on the square of the ratio of the speed to the velocity of light. (See end of Chap. IV.)

The theory of Prof. H. A. Lorentz, accordingly, shows that the shape of Michelson's stone supporting block will be distorted by the motion; its dimensions across and along the line of ether drift being affected differently. And the amount of the change will be such as precisely to compensate and neutralise the optical effect of motion which might otherwise be perceived. This theory is now generally accepted.

It is this neutralising or compensatory effect—which acts equally on to and fro motion of light, to and fro motion of electric currents, and on the shape of material bodies—that renders any positive result in experiments on ether-drift so difficult or impossible to obtain; so that, in spite of the speed with which we are rushing through space, no perceptible influence is felt on either electrical or optical phenomena, except those which are due to relative motion of source and observer.

Some Details in the Theory of the Doppler Effect, or Effect of Motion on Dispersion by Prism or Grating.

When light is analysed by a prism or grating into a spectrum, the course of each ray is de-

ABERRATION THEORY

flected—refracted or diffracted—by an amount corresponding to its frequency of vibration or wave-length.

Motion of the medium, so long as it is steady, affects neither frequency nor wave-length, and accordingly is without influence on the result. It produces no Doppler effect except when waxing or waning.

Motion of the source alone crowds the waves together on the advancing side and spreads them out on the receding side. An observer, therefore, whom the source is approaching receives shorter waves, and one from whom the source is receding receives longer waves, than normal. At any fixed point waves will arrive, therefore, with modified frequency.

So long as a source is stationary the wave-lengths emitted are quite normal, but motion of an observer may change the frequency with which they are *received*, in an obvious way; they are swept up faster if the receiver is approaching, they have a stern chase if it is receding.

All this is familiar, and was geometrically illustrated in Chapter III, but there are some minor and rather curious details which are worthy of brief consideration.

Grating Theory.

For suppose a "grating" is used to analyse the light. Its effect can depend on nothing kinetic;

it must be regulated by the merely geometric width of the ruled spaces on it. Consequently it can only directly apprehend wave-lengths, not frequencies.

In the case of a moving *source*, therefore, when the wave-length is really changed, a grating will appreciate the fact, and will show a true Doppler effect. But in the case of a moving *observer*, when all the waves received are of normal length, though swept up with abnormal frequency, the grating must still indicate wave-length alone, and accordingly will show no true Doppler effect.

But inasmuch as the telescope or line of vision is inclined at the angle of disperson to the direction of the incident ray, ordinary aberration must come in, as it always does when an observer moves athwart his line of vision; and so there will be a spurious or apparent Doppler effect due to common aberration. That is to say a spectrum line will not be seen in its true place, but will appear to be shifted by an amount almost exactly imitative of a real Doppler effect—the imitation being correct up to the second order of aberration magnitude. The slight outstanding difference between them is calculated in my *Philosophical Transactions* paper, 1893, page, 787. It is too small to observe.

It is not an important matter, but as it is rather troublesome to work out the diffraction observed by a grating advancing toward the

ABERRATION THEORY

source of light, it may be as well to record the result here.

The following are the diffracted rays which require attention—with the inclination of each to the grating-normal specified:—

The diffracted ray if all were stationary, θ_0;
The real diffracted ray when grating is advancing, ϕ;
The ray as perceived, allowing for aberration, θ;
The equivalent diffracted ray if all were stationary and the wave-length really shortened, θ_1.

As an auxiliary we use the aberration angle ϵ, such that $\sin \epsilon = a \sin \theta$, where $a = v/V$.

Among these four angles the following relations hold; so that, given one of them, all are known.

$$\begin{cases} \theta = \phi - \epsilon \\ \sin \theta_1 = (1 - a) \sin \theta_0 \\ \sin \phi = (1 - a \text{ vers } \phi) \sin \theta_0 \end{cases}$$

Whence θ and θ_1 are very nearly but not absolutely the same. θ_1 is the ray observed by an instrument depending primarily on frequency, like a prism; θ is the ray observed by an instrument depending primarily on wavelength, like a grating.

Prism Theory.

Now let a prism be used to analyse the light; its dispersive power is in most theories held to depend directly upon frequency—*i.e.*, upon a time relation between the period of a light vibration and the period of an atomic or electronic revolution or other harmonic excursion.

Let us say, therefore, that prismatic dispersion directly indicates frequency. It cannot depend upon wave-length, for the wave-length inside different substances is different, and though refractive index corresponds to this, dispersive power does not.

In the case of a prism, therefore, no distinction can be drawn between motion of source and motion of receiver; for in both cases the frequency with which the waves are received will be altered—either because they are really shorter, though arriving at normal speed, or because they are swept up faster, although of normal length.

Achromatic Prism.

It must be noticed that the observation of Doppler effect by a prism depends entirely on dispersion—*i.e.*, on waves of different length being affected differently. But prisms can be constructed whose dispersion is corrected and neutralised. Such achromatic prisms, if perfectly achromatic, will treat waves of all sizes alike; and, accordingly, the shortening of the waves from a moving source will not produce any effect. Achromatic prisms will therefore

ABERRATION THEORY

behave to terrestrial and to extra-terrestrial sources—*i.e.*, to relatively stationary and relatively moving sources, in the same way.

This must be recollected in connection with several of the negative results rightly obtained by some observers; such as Arago, for instance, who applied an achromatic prism to a star which the earth was approaching, without observing any effect. A Doppler effect should have been observed by a dispersive prism, but not by an achromatic one: for the refractive index of a substance is not affected by any motion of the earth.

It is not reasonable to expect that refractive index would be affected, since it depends in simple geometrical fashion on retarded velocity—*i.e.*, on optical etherial loading or apparent extra internal density.

An achromatic *grating*, however, is (rashly speaking) an impossibility.

Effect of Transparent Matter.

But when a ray is travelling through transparent matter, will not motion of that matter affect its course?

If the matter is moved relatively to source and receiver, as in Fizeau's experiment with running water, most certainly it will; to the full effect of the loading or extra or travelling density, (μ^2-1), compared with the total density μ^2.

THE ETHER OF SPACE

This fraction of the velocity of the material medium must directly influence the velocity of light, for the waves will be conveyed in the sense of the material motion u, with the additional speed $u\,(\mu^2-1)\,\mu^2$. (*See also* Appendix 3.)

But if the transparent matter through which the light is going is stationary with respect to source and receiver, only sharing with them the general planetary motion—*i.e.*, being subject to the opposite all-pervading ether drift—then no influence due to the drift can be experienced; for the free ether of space behaves just as it would if the matter were not there. This can be shown more elaborately by the following calculation.

Optical Effect of Ether Drift through Dense Stationary Bodies.

The calculation of the lag in phase caused by Fresnel's etherial motion may proceed thus: A dense slab of thickness z, which would naturally be traversed with the velocity V/μ, is traversed with the velocity $(V/\mu)\cos\epsilon + (v/\mu^2)\cos\theta$; where v is the relative velocity of the ether in its neighbourhood; whence the time of journey through it is

$$\frac{\mu z}{V\left(\cos\epsilon + \frac{a}{\mu}\cos\theta\right)}, \text{ instead of } \frac{\mu z}{V}$$

ABERRATION THEORY

So the equivalent air thickness, instead of being $(\mu-1)z$, is

$$\frac{\mu z}{\cos \epsilon + \frac{a}{\mu}\cos \theta} - z = \left(\frac{\mu \cos \epsilon - a \cos \theta}{\left(1 - \frac{a}{\mu}\right)^2} - 1\right) z,$$

or, to the first order of minutiæ,

$$(\mu-1)z - az \cos \theta;$$

θ being the angle between ray and ether drift inside the medium.

So the extra equivalent air layer *due to the motion* is approximately $\pm a\, z \cos \theta$, a quantity independent of μ.

Hence, no plan for detecting this first order effect of motion is in any way assisted by the use of dense stationary substances; their extra ether, being stationary, does not affect the lag caused by motion, except indeed in the second order of small quantities, as shown above.

Direct experiments made by Hoek,[1] and by Mascart, on the effect of introducing tubes of water into the path of half beams of light, are in entire accord with this negative conclusion.

Thus, then, we find that no general motion of the entire medium can be detected by changes in direction, or in frequency, or in phase; for on

---

[1] *Archives Néerlandaises* (1869), Vol. IV, p. 443, or *Nature*, Vol. XXVI, p. 500. Also Chapter IV, above.

THE ETHER OF SPACE

none of them has it any appreciable (*i.e.*, first order) effect, even when assisted by dense matter.

Another mode of stating the matter is to say that the behaviour of ether inside matter is such as to enable a potential-function,

$$\int \mu^2 v \cos \theta \, ds,$$

to exist throughout all transparent space, so far as motion of ether alone is concerned (*see* Appendix 3).

The existence of this potential function readily accounts for the absence of all effect on direction due to the general drift of the medium, whether in the presence of dense matter (such as water-filled telescopes) or otherwise. Whatever may be the path of a ray by reason of reflection or refraction in a stationary ether, it is precisely the same in a moving one if this condition is satisfied, although the wave-normals and wave-fronts are definitely shifted.

However matter affects or loads the ether inside it, it cannot on this theory be said either to hold it still, or to carry it with it. The general ether stream must remain unaffected, not only near, but inside matter, if rays are to retain precisely the same course as if it were relatively stationary.

ABERRATION THEORY

But it must be understood that the etherial motion here contemplated is the *general drift of the entire medium;* or its correlative, the uniform motion of all the matter concerned. There is nothing to be said against aberration effects being producible or modifiable by motion of *parts* of the medium, or by the artificial motion of transparent bodies and other partitioned-off regions. *Artificial* motion of matter may readily alter both the time of journey and the path of a ray, for it has no potential conditions to satisfy; it may easily describe a closed contour, and may take part in conveying light.

But I must repeat that this conveyance of light by moving matter is an effect due to the material load only; it represents no disturbance of the ether of space. Fresnel's law, in fact, definitely means that moving transparent matter does *not* appreciably disturb the ether of space. Direct experiment, as recorded in Chapter V, shows that close to rapidly moving opaque matter there is no disturbance either.

I regard the non-disturbance of the ether of space by moving matter as established.

SUMMARY.

The estimates of this book, and of *Modern Views of Electricity*, are that the ether of space is a continuous, incompressible, stationary,

THE ETHER OF SPACE

fundamental substance or perfect fluid, with what is equivalent to an inertia-coefficient of 10^{12} grammes per c.c.; that *matter* is composed of modified and electrified specks, or minute structures of ether, which are amenable to mechanical as well as to electrical force and add to the optical or electric density of the medium; and that elastic-rigidity and all potential energy are due to excessively fine-grained etherial circulation, with an intrinsic kinetic energy of the order 10^{33} ergs per cubic centimeter.

APPENDIX 1.

ON GRAVITY AND ETHERIAL TENSION

IN the arithmetical examples of Chapter IX we reckon merely the force between two bodies; but the Newtonian tension mentioned in Chapter VIII does not signify that force, but rather a certain condition or state of the medium, to variations in which, from place to place, the force is due. This Newtonian tension is a much greater quantity than the force to which it gives rise; and, moreover, it exists at every point of space, instead of being integrated all through an attracted body.

It rises to a maximum value near the surface of any spherical mass; and if the radius be R and the gravitational intensity is g, the tension at the surface is $T_0 = gR$. At any distance r, further away, the tension is $T = gR^2/r$.

This follows at once thus:—

Stating the law of gravitation as $F = \gamma \frac{m m'}{r^2}$, the meaning here adopted for etherial tension at the surface of the earth is

$$T = \int_R^\infty \frac{\gamma E}{r^2} dr = \frac{\gamma E}{R};$$

so that the ordinary intensity of gravity is

$$g = -\frac{dT}{dR} = \frac{\gamma E}{R^2} = \frac{4}{3} \pi \rho \gamma R.$$

Accordingly, near the surface of a planet the tension

THE ETHER OF SPACE

is $T_0 = gR$, or for different planets is proportional to ρR^2.

The velocity of free fall from infinity to such a planet is $\sqrt{(2 T_0)}$; the velocity of free fall from circumference to centre, assuming uniform distribution of density, is $\sqrt{(T_0)}$; and from infinity to centre it is $\sqrt{(3T_0)}$.

Expanding all this into words:—

The etherial tension near the earth's surface, required to explain gravity by its rate of variation, is of the order 6×10^{11} c.g.s. units. The tension near the sun is 2500 times as great (p. 112). With different spheres in general, it is proportional to the density and to the superficial area. Hence, near a bullet one inch in diameter, it is of the order 10^{-6}; and near an atom or an electron about 10^{-21} c.g.s.

If ever the tension rose to equal the constitutional elasticity or intrinsic kinetic energy of the ether—which we have seen is 10^{33} dynes per square centimeter (or ergs per c.c.) or 10^{22} tons weight per square millimeter—it seems likely that something would give way. But no known mass of matter is able to cause anything like such a tension.

A smaller aggregate of matter would be able to generate the velocity of light in bodies falling toward it from a great distance; and it may be doubted whether any mass so great as to be able to do even that can exist in one lump.

In order to set up a tension equal to what is here suspected of being a critical, or presumably disruptive, stress in the ether (10^{33} c.g.s.), a globe of the density of the earth would have to have a radius of eight light years. In order to generate a

GRAVITATIONAL TENSION

velocity of free fall under gravity equal to the velocity of light, a globe of the earth's density would have to be equal in radius to the distance of the earth from the sun, or say 26,000 times the earth's radius. If the density were less, the superficial area would have to be increased in proportion, so as to keep ρR^2 constant.

The whole visible universe within a parallax of $\frac{1}{1000}$ second of arc, estimated by Lord Kelvin as the equivalent of 10^9 suns, would be quite incompetent to raise etherial tension to the critical point 10^{33} c.g.s. unless it were concentrated to an absurd degree; but it could generate the velocity of light with a density comparable to that of water, if *mass* were constant.

If the average density of the above visible universe (which may be taken as 1.6×10^{-23} grammes per c.c.) continued without limit, a disruptive tension of the ether would be reached when the radius was comparable to 10^{15} light years; and the velocity of light would be generated by it when the radius was 10^7 light years. But heterogeneity would enable these values to be reached *more* easily.

Gravitation is thus supposed to be the result of a mechanical tension inherently, and perhaps instantaneously, set up throughout space whenever the etherial structure called an electric charge comes into existence; the tension being directly proportional to the square of the charge and inversely as its linear dimensions. *Cohesion* is quite different, and is due to a residual electrical attraction between groups of neutral molecules across molecular distances: a variant or modification of chemical affinity.

APPENDIX 2.

CALCULATION IN CONNECTION WITH ETHER DENSITY

JUST as the rigidity of the ether is of a purely electric character, and is not felt mechanically —since mechanically it is perfectly fluid—so its density is likewise of an electro-magnetic character, and again is not felt mechanically, because it cannot be moved by mechanical means. It is by far the most stationary body in existence; though it is endowed with high intrinsic energy of local movement, analogous to turbulence, conferring on it gyrostatic properties.

Optically, its rigidity and density are both felt, since optical disturbances are essentially electromotive. Matter loads the ether optically, in accordance with the recognised fraction $\frac{u^2-1}{u^2}$; and this loading, being part and parcel of the *matter*, of course travels with it. It is the only part amenable to mechanical force.

The mechanical density of matter is a very small portion of the etherial density; whereas the optical or electrical density of matter—being really that of ether affected by the intrinsic or constitutional electricity of matter—is not so small. The relative optical virtual density of the ether inside matter

ETHER DENSITY

is measured by μ^2; but it may be really a defect of elasticity, at least in non-magnetic materials.

Electrical and optical effects depend upon e. Mechanical or inertia effects depend upon e^2. Electric charges can load the ether optically, quite appreciably; but as regards mechanical loading, the densest matter known is trivial and gossamer-like compared with the unmodified ether in the same space.

Massiveness of the Ether deduced from Electrical Principles.

Each electron, moving like a sphere through a fluid, has a certain mass associated with it; dependent on its size, and, at very high speeds, on its velocity also.

If we treat the electron merely as a sphere moving through a perfect liquid, its behaviour is exactly as if its mass were increased by half that of the fluid displaced and the surrounding fluid were annihilated.

Ether being incompressible, the density of fluid inside and outside an electron must be the same. So, dealing with it in this simplest fashion, the resultant inertia is half as great again as that of the volume of fluid corresponding to the electron: that is to say the effective mass is $2\pi\rho a^3$, where ρ is the uniform density. If an electron is of some other shape than a sphere, then the numerical part is modified, but remains of the same order of magnitude, so long as there are no sharp edges.

If, however, we consider the moving electron as

THE ETHER OF SPACE

generating circular lines of magnetic induction, by reason of some rotational property of the ether, and if we attribute all the magnetic inertia to the magnetic whirl thus caused round its path—provisionally treating this whirl as an actual circulation of fluid excited by the locomotion—then we shall proceed thus:—

Let a spherical electron e of radius a be flying at moderate speed u, so that the magnetic field at any point, $r\theta$, outside, is

$$H = \frac{eu \sin\theta}{r^2},$$

and the energy per unit volume everywhere is $\mu H^2/8\pi$.

But a magnetic field has been thought of by many mathematicians as a circulation of fluid along the lines of magnetic induction—which are always closed curves—at some unknown velocity w.

So consider the energy per unit volume anywhere: it can be represented by the equivalent expressions

$$\tfrac{1}{2}\rho w^2 = \frac{\mu H^2}{8\pi} = \frac{\mu}{8\pi} \cdot \frac{e^2 u^2 \sin^2\theta}{r^2};$$

wherefore

$$\frac{w}{u} = \sqrt{\left(\frac{\mu}{4\pi\rho}\right)} \cdot \frac{e \sin\theta}{r^2}.$$

The velocity of the hypothetical circulation must be a maximum at the equator of the sphere, where $r = a$ and $\theta = 90$; so, calling this w_0,

$$\frac{w_0}{u} = \sqrt{\left(\frac{\mu}{4\pi\rho}\right)} \frac{e}{a^2},$$

and
$$\frac{w}{w_0} = \frac{a^2 \sin\theta}{r^2};$$

wherefore the major part of the circulation is limited to a region not far removed from the surface of the electron.

The energy of this motion is

$$\tfrac{1}{2}\rho \int_0^\pi \int_a^\infty w^2 \cdot 2\pi r \sin\theta \cdot r d\theta \cdot dr,$$

whence, substituting the above value of w, the energy comes out equal to $\tfrac{4}{3}\pi\rho a^3 w_0^2$.

Comparing this with a mass moving with speed u,

$$m = \frac{8}{3}\pi\rho a^3 \left(\frac{w_0}{u}\right)^2.$$

This agrees with the simple hydrodynamic estimate of effective inertia if $w_0 = \tfrac{1}{2}\sqrt{3}.u$; that is to say, if the whirl in contact with the equator of the sphere is of the same order of magnitude as the velocity of the sphere.

Now for the real relation between w_0 and u we must make a hypothesis. If the two are considered equal, the effectively disturbed mass comes out as twice that of the bulk of the electron. If w_0 is smaller than u, then the mass of the effectively disturbed fluid is less even than the bulk of an electron; and in that case the estimate of the fluid-density ρ must be *exaggerated* in order to supply the required energy. It is difficult to suppose the equatorial circulation w_0 *greater* than u, since it is generated by it; and it is most reasonable to treat them both as of the same order of magnitude. So, taking them as equal,

$$e \rightleftharpoons a^2\sqrt{\frac{4\pi\rho}{\mu}}$$

and $m =$ twice the spherical mass.

Hence all the estimates of the effective inertia of an electron are of the same order of magnitude, being all comparable with that of a mass of ether equal to the electron in bulk. But the linear dimension of an electron is 10^{-13} centimeter diameter, and its mass is of the order 10^{-27} gramme. Consequently the density of its material must be of the order 10^{12} grammes per cubic centimeter.

This, truly, is enormous, but any reduction in the estimate of the circulation-speed, below that of an electron, would only go to increase it. And, since electrons move sometimes at a speed not far below that of light, we cannot be accused of underestimating the probable velocity of magnetic spin by treating it as of the same order of magnitude, at the bounding surface of the electron, as its own speed: a relation suggested, though not enforced, by gyrostatic analogies.

Some Consequences of this Great Density.

The amplitude of a wave of light, in a place where it is most intense, namely near the sun where its energy amounts to 2 ergs per c.c., comes out only about 10^{-17} of the wave-length. The maximum

ETHER DENSITY

tangential stress called out by such strain is of the order 10^{11} atmospheres.

The hypothetical luminous circulation-velocity, conferring momentum on a wave-front, in accordance with Poynting's investigation, comes out 10^{22} cm. per sec. These calculations are given in the concluding chapter of the new edition of *Modern Views of Electricity*.

The supposed magnetic ethereal drift, along the axis of a solenoid or other magnetic field, if it exist, is comparable to .003 centim. per sec., or 4 inches an hour, for a field of intensity 12,000 c.g.s.

But it is not to be supposed that this hypothetical velocity is slow everywhere. Close to an electron the speed of magnetic drift is comparable to the locomotion-velocity of the electron itself, and may therefore rise to something near the speed of light; say $\frac{1}{30}$ th of that speed: but in spite of that, at a distance of only 1 millimeter away, it is reduced to practical stagnation, being less than a millimicron per century.

In any solenoid, the ampere-turns per linear inch furnish a measure of the speed of the supposed magnetic circulation along the axis—no matter what the material of the core may be—in millimicrons per sec.

[1 micron $= 10^{-6}$ meter; 1 millimicron is 10^{-9} meter $= 10^{-7}$ centimeter, or a millionth of a millimeter.]

To get up an ethereal speed of 1 centimeter per second—such as might be detected experimentally by refined optical appliances, through its effect in

THE ETHER OF SPACE

accelerating or retarding the speed of light sent along the lines of magnetic force—would need a solenoid of great length, round every centimeter of which 1000 amperes circulated 3000 times. That is to say, a long field of four million c.g.s. units of intensity.

In other words, any streaming along magnetic lines of force, such as could account for the energy of a magnetic field, must be comparable, in centimeters per second, to one four-millionth of the number of c.g.s. units of intensity in the magnetic field.

APPENDIX 3.

FRESNEL'S LAW A SPECIAL CASE OF A UNIVERSAL POTENTIAL FUNCTION

THE modern view of Fresnel's Law may be worded thus:—

Inside a region occupied by matter, in addition to the universal ether of space, are certain modified or electrified specks, which build up the material atoms. These charged particles, when they move, have specific inertia, due to the magnetic field surrounding each of them. And by reason of this property, and as a consequence of their discontinuity, they virtually increase the optical density of the ether of space, acting in analogy with weights distributed along a flexible cord. Thus they reduce the velocity of light in the ratio of the refractive index $\mu : 1$, and therefore may be taken as increasing the virtual density of the ether in the ratio $1 : \mu^2$.

That is to say, their loading makes the ether behave to optical waves as if—being a homogeneous medium without these discontinuous loads—it had a density μ^2 times that which it has in space outside matter. Calling the density outside 1, the extra density inside must be $\mu^2 - 1$, so as to make up the total to μ^2.

THE ETHER OF SPACE

The μ^2-1 portion is that which we call "matter," and this portion is readily susceptible to locomotion, being subject to—that is, accelerated by—mechanical force. The free portion of normal density 1 is absolutely stationary as regards locomotion, whether it be inside or outside a region occupied by ordinary matter, for it is not amenable to either mechanical or electric forces. They are transmitted by it, but never terminate upon it; except, indeed, at the peculiar structure called a wave-front, which simulates some of the properties of matter.

(If free or unmodified ether can ever be moved at all, it must be by means of a magnetic field; along the lines of which it has, in several theories, been supposed to circulate. Even this, however, is not real locomotion.)

Fizeau tested that straightforward consequence of this theory which is known as Fresnel's law, and ascertained by experiment that a beam of light was accelerated or retarded by a stream of water, according as it travelled with or against the stream. And he found the magnitude of the effect precisely in accordance with the ratio of the locomotive portion of the ether to the whole—the fraction $(\mu^2 - 1)/\mu^2$ of the speed of the water being added to or subtracted from the velocity of light, when a beam was sent down or up the stream.

But even if another mode of expression be adopted, the result to be anticipated from this experiment would be the same.

For instead of saying that a modified portion of the ether is moving with the full velocity of the

FRESNEL'S LAW

body while the rest is stationary, it is permissible for some purposes to treat the whole internal ether as moving with a fraction of the velocity of the body.

On this method of statement the ether outside a moving body is still absolutely stationary, but, as the body advances, ether may be thought of as continually condensing in front, and, as it were, evaporating behind; while, inside, it is streaming through the body in its condensed condition at a pace such that what is equivalent to the normal quantity of ether in space may remain absolutely stationary. To this end its speed backward relatively to the body must be u/μ^2 and accordingly its speed forward in space must be $u(1 - 1/\mu^2)$.

For consider a slab of matter moving flatways with velocity u; let its internal etherial density be μ^2, and let the external ether of density 1 be stationary. Let the forward speed of the internal ether through space be xu, so that a beam of light therein would be hurried forward with this velocity. Then consider two imaginary parallel planes moving with the slab, one in advance of it and the other inside it, and express the fact that the amount of ether between those two planes must continue constant. The amount streaming relatively backward through the first plane as it moves will be measured by u times the external density, while the amount similarly streaming backward through the second plane will be $(u-xu)$ times the internal density. But this latter amount must equal the former amount. In other words,

$$u \times 1 \text{ must equal } (u - xu) \times \mu^2.$$

THE ETHER OF SPACE

Consequently x comes out $x = (\mu^2 - 1)/\mu^2$; which is Fresnel's incontrovertible law for the convective effect of moving transparent matter on light inside it.

The whole subject, however, may be treated more generally, and for every direction of the ray, on the lines of Chapter X, thus:—

Inside a transparent body light travels at a speed V/μ; and the ether, which outside drifts at velocity v, making an angle θ with the ray, inside may be drifting with velocity v' and angle θ'.

Hence the equation to a ray inside such matter is

$$T' = \int \frac{ds}{(V/\mu)\cos\epsilon' + v'\cos\theta'} = \min.,$$

where $\dfrac{\sin\epsilon'}{\sin\theta'} = \dfrac{v'}{V/\mu} = a'.$

This may be written

$$T' = \int \frac{\cos\epsilon'\, ds}{V/\mu\,(1-a'^2)} - \int \frac{v'\cos\theta'\, ds}{V^2/\mu^2\,(1-a'^2)};$$

the second term alone involves the first power of the motion, and assuming that $\mu^2 v' \cos\theta' = d\phi'/ds$, and treating a' as a quantity too small for its possible variations to need attention, the expression becomes

$$T' = \mu T \frac{\cos\epsilon'}{1-a'^2} - \frac{\phi'_B - \phi'_A}{V^2(1-a'^2)},$$

T being the time of travel through the same space when empty. Now, if the time of journey and course of ray, however they be affected by the dense

166

FRESNEL'S LAW

body, are not to be *more* affected by reason of etherial drift through it than if it were so much empty space, it is necessary that the difference of potential between two points A and B should be the same whether the space between is filled with dense matter or not (or, say, whether the ray-path is taken through or outside a portion of dense medium). In other words (calling ϕ the outside and ϕ' the inside potential function), in order to secure that T' shall not differ from μT by anything depending on the first power of motion, it is necessary that ϕ'B $-\phi'$A shall equal ϕB $-\phi$A; *i.e.*, that the potential inside and outside matter shall be the same up to a constant, or that $\mu^2 v' \cos \theta' = v \cos \theta$; which for the case of drift along a ray is precisely Fresnel's hypothesis.

Another way of putting the matter is to say that to the first power of drift velocity

$$T' = \mu T - \int (\mu^2 v' \cos \theta' - v \cos \theta) \, ds / V^2,$$

and that the second or disturbing term must vanish.

Hence Fresnel's hypothesis as to the behaviour of ether inside matter is equivalent to the assumption that a potential function, $\int \mu^2 v \cos \theta \, ds$, exists throughout all transparent space, so far as motion of ether alone is concerned.

Given that condition, no first-order interference effect due to drift can be obtained from stationary matter by sending rays round any kind of closed contour; nor can the path of a ray be altered by

THE ETHER OF SPACE

etherial drift through any stationary matter. Hence filling a telescope tube with water cannot modify the observed amount of stellar aberration.

The equation to a ray in transparent matter moving with velocity u in a direction ϕ, and subject to an independent ether drift of speed v in direction θ, is

$$\int \frac{ds}{V/\mu \cos \epsilon + v/\mu^2 \cos \theta + u[1 - (1/\mu^2)]\cos \phi} = \text{const.}$$

THE END

Made in the USA
Lexington, KY
14 January 2011